Combating the Silent Threat: Internal Corrosion Control in Oil & Gas Pipelines

Jacks

TABLE OF CONTENTS

Page

CHAPTER 1: INTRODUCTION

History and Background

Corrosion is the gradual degradation of metals due to their interaction with the environment and corrosion of metals is unavoidable.[1] Corrosion causes major health, safety, and environmental problems. The estimated global cost of corrosion is US \$2.5 trillion/year, which is roughly equal to 3.4% of the gross domestic product (GDP). It has been reported that the cost related to corrosion of oil and gas pipelines in the United States is roughly \$7 billion annually.[2] Both external and internal corrosion processes occur in oil and gas pipelines. The scope of this research lies in retarding of internal corrosion of oil and gas transportation pipelines. The most common methods used to mitigate corrosion are corrosion resistant alloys,[3] protective coatings,[4] cathodic protection,[5] and corrosion inhibitors.[6, 7] Among these, protective coatings and cathodic potential are suitable for external corrosion protection of underground pipelines and storage tanks. The use of corrosion-resistant alloys is very expensive which requires high initial costs, making it not a viable option. The application of corrosion inhibitors is an efficient and economical way of retarding internal corrosion upon adding a small concentration (ppm) in a fluid stream.

Oil and gas transportation pipelines are made of low carbon steel, corroded internally when continuously in contact with H_2O, CO_2, and H_2S.[8-10] A pure carbon steel in contact with adjacent liquids and gases could be modified to $FeCO_3$, FeS, Fe oxides, and Fe hydroxides. CO_2 is one of the main corrosive substances in oil and gas transportation pipelines. The general mechanism of CO_2 corrosion (sweet corrosion) is presented below.

Iron atoms from the metal surface lose electrons at anodes, which are taken by reducible species such as H^+ at cathodes.[8, 11]

The CO_2 gas can dissolve in water:

$$CO_{2(g)} \rightleftharpoons CO_{2(aq)} \tag{1}$$

Small fraction of dissolved CO_2 form carbonic acid:

$$CO_{2(aq)} + H_2O_{(aq)} \rightleftharpoons H_2CO_{3(aq)} \tag{2}$$

Carbonic acid partially dissociates into hydrogen ions and bicarbonate ions, then further dissociate to form additional hydrogen ions and carbonate ions:

$$H_2CO_{3(aq)} \rightleftharpoons HCO_3^-{}_{(aq)} + H^+{}_{(aq)} \tag{3}$$

$$HCO_3^-{}_{(aq)} \rightleftharpoons CO_3^{2-}{}_{(aq)} + H_{2(g)} \tag{4}$$

Hydrogen ions are reduced to form hydrogen gas, which is the main cathodic reaction associated with CO_2 corrosion:

$$2H^+{}_{(aq)} + 2e^- \rightarrow H_{2(g)} \tag{5}$$

Carbonic acid can be directly reduced at the metal surface.[12, 13]

$$2H_2CO_3 + 2e^- \rightarrow H_{2(aq)} + 2HCO_3^-{}_{(aq)} \tag{6}$$

While oxidation of surface iron results in steel corrosion:

$$Fe_{(s)} \rightarrow Fe^{2+}{}_{(aq)} + 2e^- \tag{7}$$

Overall reaction:

$$CO_{2(g)} + H_2O_{(aq)} + Fe_{(s)} \rightarrow FeCO_{3(s)} + H_{2(g)} \tag{8}$$

Organic inhibitors are the most widely used inhibitors for corrosion control of oil and gas pipeline industries because of their low toxicity, solubility, high hydrocarbon content, high efficacy, and the ability to adsorb onto metals to mitigate corrosion.[14]

Corrosion inhibitors are amphiphilic molecules that contain a hydrophilic head group and a hydrophobic chain group. A strong association between the metal surface and the polar head group is favorable whereas a high concentration of hydrocarbon chain group and corrosive substance is not favorable. These properties allow surfactants as good candidates for corrosion inhibitors.[15] This research focuses only on Quaternary ammonium, imidazoline, and alkyl trimethylammonium-based surfactants (**Figure 1.1**). Five different chain lengths of Quat 1 and imidazoline were synthesized, and Quat 2 was commercially available. Self-assembled monolayer formation of these surfactants was investigated at air-liquid and solid-liquid interfaces using sum frequency generation spectroscopy.

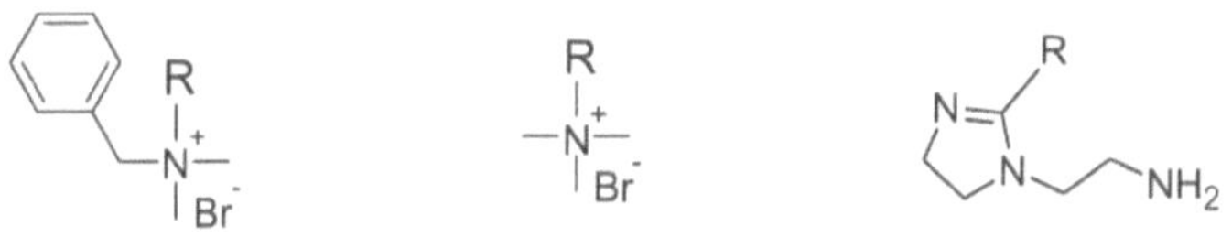

(a)   (b)   (c)

Figure 1.1: Three distinct types of surfactants a) alkyl benzyldimethylammonium bromides (Quat 1), b) alkyl trimethylammonium bromides (Quat 2), and c) alkyl imidazolines was considered. R represents linear alkyl chains, $(CH_2)nCH_3$, where n = 3, 5, 7, 9, and 11.

So far, the inhibition mechanism of surfactants was studied by saturation adsorption data,[6] electrochemical impedance spectroscopy (EIS),[16-18] second harmonic generation laser scattering,[18] and atomic force microscopy (AFM).[19] The results of the previous experiments have shown inhibition efficiency of the corrosion inhibitors is related to the surface coverage and the corrosion rate decreases as the amount of adsorbed surfactant increases.[20] Second finding is the structure of the hydrophilic head group plays a significant role in surfactant adsorption.[18, 21] Lastly, the alkyl chain length governs and proportional to the inhibitor efficiency.[18, 22, 23] These previous results lead to the investigation of fundamental adsorption behavior of surfactants using SFG spectroscopy; a powerful tool for surface and interface study.

Sum Frequency Generation (SFG) Spectroscopy

Sum frequency generation (SFG) spectroscopy is a second-order nonlinear optical technique that obtains the vibrational spectra of the interfacial molecules.[24] The analysis of the vibrational spectra provides molecular conformation, polar orientation, and average tilt angle of the adsorbed molecules to the surface normal.[25-27] The sum frequency (SF) signal intensity from adsorbed molecules resonantly enhance when the incident infrared beam frequency matches with a vibrational mode at the interface. SF beam can be plotted as a function of infrared frequency; a vibrational spectrum is obtained. For a molecular vibrational mode to be SF active, it must be in an asymmetric environment on both macroscopic and microscopic scales. The bulk phase is centrosymmetric due to an isotropic distribution of molecules, and therefore no SF signal arises from bulk. However, this symmetry breaks at the interface and interfacial molecules are in an

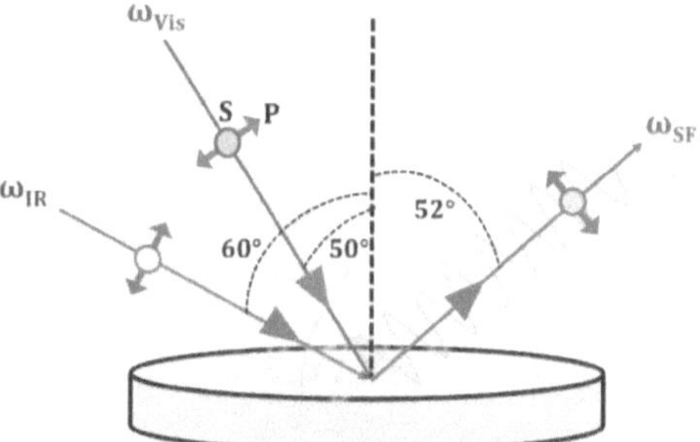

Figure 1.2: The visible beam (795 nm) and IR beam overlap at the sample surface at an incident angle 50° and 52° respectively to the surface normal and generates SF beam 52° from the surface normal.[28]

asymmetric environment thus will be SF active. For the SFG process, two pulsed laser beams; visible beam (ω_{VIS}) and tunable infrared beam (ω_{IR}), achieve temporal and spatial overlap at the sample surface, the generated beam is the sum of the frequencies of two beams known as SFG beam.

$$\omega_{SF} = \omega_{VIS} + \omega_{IR} \tag{9}$$

The SFG signal intensity is directly proportional to the square of second-order nonlinear susceptibility, $\chi^{(2)}$.

$$I(\omega_{SF}) \propto \left[\sum_i \chi_{eff}^{(2)} \right]^2 \tag{10}$$

Where i, j, and k coordinates indicate x, y, z-axes, respectively. $\chi_{eff}^{(2)}$ is a third rank tensor and can be defined by using Fresnel factors and macroscopic nonlinear susceptibilities, $\chi_{ijk}^{(2)}$, at the interface, as shown in equation 11.

$$\chi^{(2)}_{\text{eff},ijk} = \chi^{(2)}_{ijk} [L_i e_i][L_j e_j][L_k e_k] \qquad (11)$$

Where L represents Fresnel factor and e represents the unit optical field vector.

Macroscopic susceptibility $\chi^{(2)}_{ijk}$ has 27 components, However, due to symmetry

constraints, some of the components become almost negligible. $\chi^{(2)}_{\text{eff}}$ is composed of (1)

the resonant susceptibility, $\chi^{(2)}_R$, and (2) the nonresonant susceptibility, $\chi^{(2)}_{NR}$ (equation

12). The resonant term, $\chi^{(2)}_R$, as shown in equation 12, is due to vibrational transitions of

the adsorbed surface molecules. Whereas the nonresonant term, $\chi^{(2)}_{NR}$, originates from the

surface plasmon resonance and invariant with IR frequency. $\chi^{(2)}_{NR}$ is very small for

dielectric and liquid surfaces.

$$\chi^{(2)}_{\text{eff}} = \chi^{(2)}_R + \chi^{(2)}_{NR} \qquad (12)$$

$$\chi^{(2)}_R = \frac{N\beta^{(2)}}{\omega_{IR} - \omega_q + i\Gamma_q} \qquad (13)$$

Where N represents the number density of vibrational modes, and $\beta^{(2)}$ represents the

orientation averaged hyperpolarizability. Hyperpolarizability is the product of Raman and

IR transition moments. This gives rise to the selection rule that SFG vibrational modes

must be both IR and Raman active. ω_{IR}, ω_q, and Γ_q^{-1} are the IR frequency, the

frequency of the qth vibrational mode, and the relaxation time.[24]

SFG spectroscopy can determine the orientation of the functional group of interfacial

adsorbed molecules.[28, 29] The orientational analysis is the determination of tilt angle to the

surface normal of the molecular group. The tilt angle was calculated by taking the

intensity ratio of different polarization of SF, visible, and IR beams. From the four

polarization combinations, only ssp and ppp polarization combinations were considered for tilt angle calculation. These combinations are in decreasing order of energy of beams. Here, 's' and 'p' are the polarized beams perpendicular and parallel to the plane of incidence. The three letters represent the three beams either s polarized or p polarized. The s polarized beam contains only y component of the optical field. On the other hand, the p polarized light contains x and z components of the optical field. Eight polarization combination is possible SFG measurements. The polarization combination and the associated susceptibilities are listed below.

sss = 0

ssp = yyz

sps = yzy

spp = 0

pss = zyy

psp = 0

pps = 0

ppp = xxz + xzx + zxx + zzz

Each polarization combination are correlated with specific susceptibility components $\chi_{ijk}^{(2)}$. As mentioned earlier, the susceptibility $\chi_{ijk}^{(2)}$ has 27 components in Cartesian space. By considering symmetry constraints, only seven components correspond to four possible polarization combinations; ssp, ppp, pss, and sps are listed in **Table 1.1**.

Table 1.1: Polarization combinations and the susceptibility elements $\chi_{ijk}^{(2)}$ that contribute to the SFG spectrum.[25]

Polarization combination	Elements of $\chi_{ijk}^{(2)}$
ssp	$\chi_{yyz}^{(2)}$
ppp	$\chi_{xxz}^{(2)},\ \chi_{zzz}^{(2)},\ \chi_{xzx}^{(2)},\ \chi_{zxx}^{(2)}$
pss	$\chi_{zyy}^{(2)}$
sps	$\chi_{yzy}^{(2)}$

The effective susceptibility can be expressed with polarization combination and of Fresnel factors are shown in equation 14

$$\chi_{eff\,ssp}^{(2)} = L_{ysfg}L_{yvis}L_{zir}\chi_{eff\,yyz}^{(2)}$$

$$\chi_{eff\,sps}^{(2)} = L_{ysfg}L_{zvis}L_{yir}\chi_{eff\,yzy}^{(2)} \qquad (14)$$

$$\chi_{eff\,pss}^{(2)} = L_{zsfg}L_{yvis}L_{yir}\chi_{eff\,zyy}^{(2)}$$

$$\chi_{eff\,ppp}^{(2)} = - L_{xsfg}L_{xvis}L_{xir}\chi_{eff\,xxz}^{(2)} - L_{xsfg}L_{zvis}L_{xir}\chi_{eff\,xzx}^{(2)} + L_{zsfg}L_{xvis}L_{xir}\chi_{eff\,zxx}^{(2)} + L_{zsfg}L_{zvis}L_{zir}\chi_{eff\,zzz}^{(2)}$$

The negative sign for the SFG Fresnel factor along x, L_{xsfg} indicates the optical field vector of x component is in opposite direction with respect to the incident field. The surface bound Cartesian coordinates (x,y,z) and the molecular bound coordinate (a,b,c) system rarely coincide. These two coordinates can be correlated with Euler transformations to establish the orientation of the surface molecules.

$$\chi_{ijk}^{(2)} = \sum_{l,m,n} \mathbb{R}_{il} \mathbb{R}_{il} \mathbb{R}_{il} \mathbb{R}_{il} \, \mathbb{R}\beta_{lmn} \qquad (15)$$

Where a,b,c are replaced by l,m,n, respectively. l, m, and n represents SFG, visible, and IR

in the molecular coordinate system.

The Euler transformations matrix $\mathbb{R}$ can be expressed as

$$\mathbb{R} = \begin{bmatrix} Cos\psi Cos\phi - Cos\theta Sin\phi Sin\psi & -Sin\psi Cos\phi - Cos\theta Sin\phi Cos\psi & Sin\theta Sin\phi \\ Cos\psi Sin\phi + Cos\theta Cos\phi Sin\psi & -Sin\psi Sin\phi + Cos\theta Cos\phi Cos\psi & -Sin\theta Cos\phi \\ Sin\theta Sin\psi & Sin\theta Cos\psi & Cos\theta \end{bmatrix}$$

(16)

Where ϕ, θ, and ψ represent the rotation, tilt, and twist angles, respectively.

β_{lmn} in equation 15 is the second-order molecular hyperpolarizability; the third rank

tensor contains 27 components. Due to the symmetry constraints, only 11 non-vanishing

elements are possible are listed in **Table1.2**.

Table 1.2: The second-order hyperpolarizability components associated with different

vibrational modes. [28]

Second-order molecular hyperpolarizability, β_{lmn}	Vibrational mode
$\beta_{aac} = \beta_{bbc},\ \beta_{ccc}$	Symmetric stretching
$\beta_{aaa} = -\,\beta_{bba} = -\,\beta_{abb} = -\,\beta_{bab}$	Asymmetric stretching
$\beta_{aca} = \beta_{caa} = \beta_{bcb} = \beta_{cbb}$	Asymmetric stretching

For example, in C$_{3v}$ point group such as for the methyl group the symmetric stretching mode $\chi_{yyz}^{(2)}$ can be obtained from the hyperpolarizability combinations.

$$\chi_{yyz,ss}^{(2)} = \beta_{ccc}\cos^2\phi\left(\left(1+\left(\frac{Sin\phi}{Cos\phi}\right)^2 R\right)Cos\theta - (1-R)Cos^3\theta\right)^2 \quad (17)$$

and $R = \dfrac{\beta_{aac}}{\beta_{ccc}}$ $\quad$ (18)

Where R is the depolarization ratio obtained from the Raman scattering and the R values for the methyl group ranges from 1.66 - 4.00. The simulated curve containing orientational tilt angle and Raman depolarization ratio R can be determined using equation 17.

From the fitted data, the intensity ratios of active vibrational modes were calculated. Then, the ratios were compared with the SFG simulated curve. **Figure 1.3** and **Figure 1.4** show the simulated curve for the SFG intensity as a function of the tilt angle and tilt angle θ for the terminal methyl group.

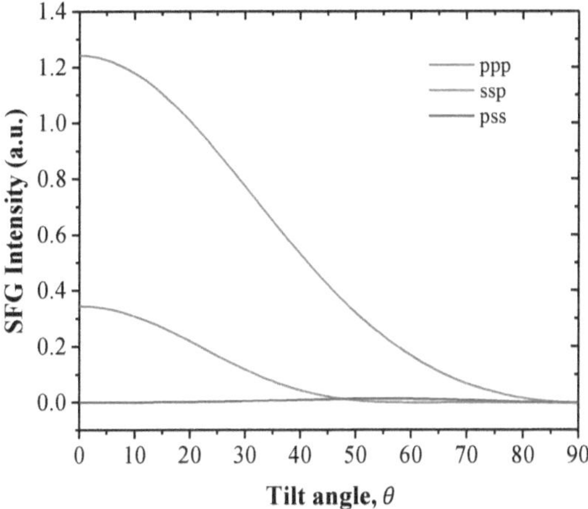

Figure 1.3: SFG simulated curve for SFG intensity vs tilt angle for the symmetric stretch vibrational mode of the methyl group.

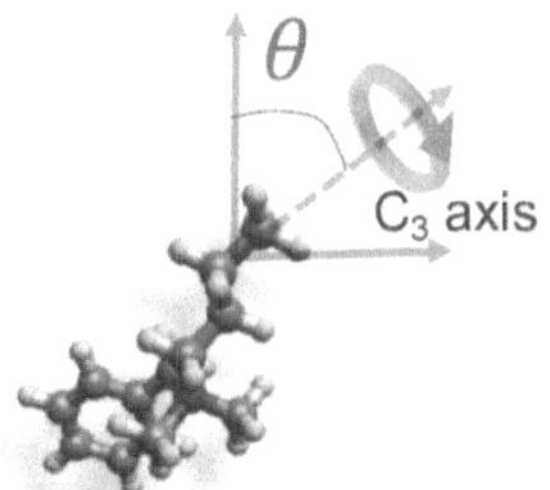

Figure 1.4: The surface and molecular bound axis systems are related by a Euler angle of θ.

The tilt angle of the terminal methyl group was estimated from the intersection point of SFG intensity ratios from the fitted data and the SFG simulated curve.

Objective and Significance

Corrosion is a phenomenon described as the irreversible damage or destruction of metals due to chemical or electrochemical reactions wherein metal atoms oxidize upon coming in contact with H_2O, CO_2, and H_2S.[8, 30] Corrosion causes economic damage to home appliances, automobiles, drinking water systems, oil and gas pipelines, bridges, and public buildings. In oil and gas transportation pipelines, corrosion is taken into account as a major concern influencing the durability and reliability of pipelines in terms of transportation of crucial energy sources in the United States. Over 25% of corrosion-related failures are experienced in the oil and gas industries in which half of these failures are related to transportation pipelines. Many corrosion prevention methods are discussed earlier. This research work focuses on studying the adsorption behavior of corrosion inhibitors which can be applied to prevent internal corrosion of oil and gas transportation pipelines. The application of inhibitors to mitigate pipelines corrosion is one of the most efficient, economical, and environmentally friendly methods.[6, 9] Several approaches are available to mitigate corrosion processes. An example of it is surfactants which can inhibit corrosion significantly by forming a chemically bonded film on the metal surface. This alters the nature of the electrochemical reaction.[16] They have been applied as corrosion inhibitors in many industrial processes; in oil and gas pipelines, alloy industries, and cooling water systems.[10] Organic inhibitors are generally considered more environmentally friendly and preferable than most other inorganic inhibitors.[31] The proposed corrosion inhibitors are Quaternary ammonium salts, imidazoline, and alkyl trimethylammonium-based surfactants because they are less toxic and highly efficient.[16, 32-34] However, the inhibition property of these surfactants is limited by various operating conditions such as temperature, flow rate, water/oil property, and the nature of the metal surface.[22, 33, 35, 36]

This research proposal focuses on the synthesis of surfactants, and investigation of the effect of structural changes of surfactants at air-liquid and solid-liquid interfaces. Then correlate what we learn from the consequences of structural changes to its effect on the adsorption behavior on air-liquid, and solid-liquid interfaces. After that, we can design better surfactants for a specific environment under specific conditions. Surfactants adsorb onto the metal surface and mitigate corrosion. However, their exact mechanism is not well understood, which leads to finding new, useful, environmentally friendly surfactant molecules.[9] The interfacial properties of proposed surfactants were investigated by varying the surfactant tail groups and headgroups. The conformational changes and orientational analysis at different interfaces will help us to correlate the adsorption behavior of these surfactants. The extent of conformational changes with increasing chain length was carried out to understand the appropriate size of the alkyl chain length which governs the corrosion inhibition efficiency. The crude oil-water chemistry can also affect the performance of corrosion inhibitors. Surfactants in aqueous solutions with high salt concentrations or ionic strength were studied to investigate the effect of salt on the interfacial conformation. This characterization provides new insight in understanding the mode of action of the surfactants, by sequentially studying the surfactants through headgroups, different chain lengths, and in the different external environment.

The primary objective of this research is to investigate the conformational changes of interfacial molecules due to the structural changes of surfactants at air-liquid and liquid-solid interfaces by sum frequency generation (SFG) spectroscopy. To achieve this main goal several specific goals were considered. In the first specific goal, alkyl benzyl dimethylammonium salt (Quats) and imidazoline-based surfactants were synthesized and characterized using ^{1}H-NMR and ^{13}C-NMR. In the second specific goal, the critical micelle concentration (CMC) values of Quats were determined by the Surface Tension method. In the third specific aim, the effect of chain length, and headgroup of Quats at different ionic strengths on the interfacial molecular conformation were investigated at the air-water interface using SFG spectroscopy. Finally, the *in-situ* adsorption of Quats and their self-assembly were investigated at a liquid-metal interface by SFG spectroscopy.

In general, SFG spectroscopy is effectively employed to identify distinct functional groups present at air-liquid, air-solid, and liquid-solid interfaces. [37-39] SFG is a useful tool to investigate bonding mechanisms, vibrational states, orientations of molecules. It has also been applied to chromatographic materials,[40] combustion environments,[41] and in tribology.[42] This technique has been successfully applied to the conformational analysis of surfactant molecules, to investigate the behavior of hydrocarbon chains adsorbed at air-liquid[43, 44] and liquid-solid interfaces.[45, 46] In this research, the SFG spectroscopy technique was employed to investigate the arrangement of surfactants at air-liquid and liquid-metal interfaces.

CHAPTER 2: SYNTHESIS OF SURFACTANTS

Synthesis of Quaternary Ammonium Bromides (Quats)

Alkylbenzyldimethylammonium bromides with five different alkyl chain lengths (butyl, hexyl, octyl, decyl, and dodecyl) have been synthesized by established methods.[47] N, N-dimethylbenzylamine (Fisher Scientific, New Hampshire, US) as a tertiary amine in acetonitrile (Fisher Scientific, New Hampshire, US) and appropriate bromoalkane (Fisher Scientific, New Hampshire, US) was reacted with the equivalent stoichiometric amount. Acetonitrile solvent was used to achieve a high rate of amine Quaternization, and bromoalkane was used by considering the cost, availability, and reactivity over other alkyl halides.[48] The glasswares required for the synthesis of Quats were acid-cleaned (mixture of concentrated H_2SO_4:HNO_3 at 3:1 by volume) overnight, then rinsed and boiled with deionized water. The clean glasswares were placed in a drying oven overnight. The synthetic procedure involves three steps; 1) reaction between tertiary amine and bromoalkane, 2) solvent removal, and 3) purification. ^1H NMR spectra of starting materials used for the synthesis were recorded as well.

The reaction parameters are tabulated in **Table 2.1**, where C4, C6, C8, C10, and C12 represent a corresponding alkyl group of that Quat compounds.

Table 2.1: Quat compounds with reaction parameters.[49]

Reactants		Solvent	Reflux	Duration	Products
Tertiary amine	Bromoalkanes		temperature		(Quats)
N, N Dimethylbenzylamine ($C_9H_{13}N$)	Bromobutane (C_4H_9Br)	Acetonitrile (ACN)	82 °C	24 hours	C4
	Bromohexane ($C_6H_{13}Br$)				C6
	Bromooctane ($C_8H_{17}Br$)				C8
	Bromodecane ($C_{10}H_{21}Br$)				C10
	Bromododecane ($C_{12}H_{25}Br$)				C12

A typical example of benzyl dimethyl butylammonium bromide synthesis is shown in **Figure 2.1**.

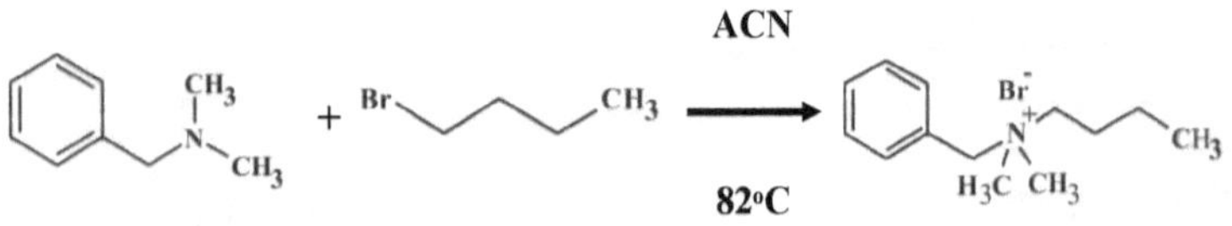

Figure 2.1: Synthesis of benzyl dimethyl butylammonium bromide

A typical example of benzyl dimethyl octylammonium bromide ^{1}H-NMR spectrum is shown in **Figure 2.2**.[49, 50]

^{1}H NMR in CDCl$_3$: δ = 7.65 (Ph, m, 2H), 7.42 (Ph, m, 3H, J = 7.1 Hz), 5.09 (Ph-CH$_2$, s, 2H), 3.54 (NCH$_2$, t, 2H, J = 8.2 Hz), 3.29 (N-CH$_3$, s, 6H), 1.77 (CH$_2$, s, 2H), 1.29 (CH$_2$, d, 14H, J = 14 Hz), 0.84 (C-CH$_3$), t, 3H, J = 6.6 Hz

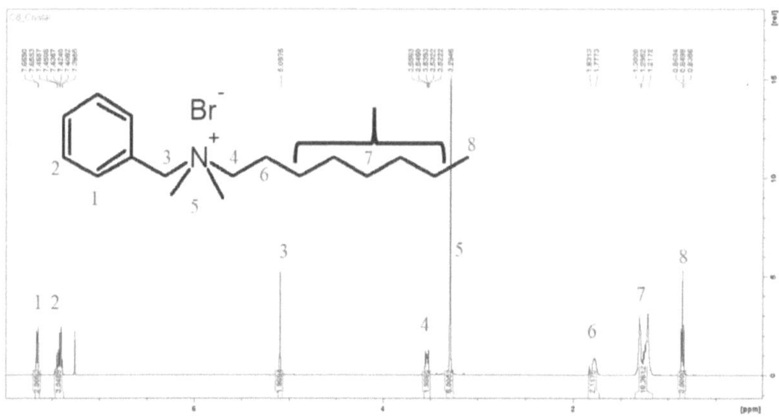

Figure 2.2. The ^{1}H-NMR spectrum of benzyldimethyloctylammonium bromide (C8).[50]

Synthesis of Benzyldimethylbutyllammonium Bromide (C4)

0.100 mole N, N-Dimethylbenzylamine and 100 ml acetonitrile were added into the three-necked round-bottom flask. Then, a condenser, an addition funnel, and a thermometer were attached to the three necks of the flask. Water flow in the condenser was ensured to reflux acetonitrile and amine mixture. Then, the round-bottom flask was heated to the reflux temperature of 82°C by using a heating mantle. After refluxing commenced, the respective 0.100 moles (equimolar) bromobutane was added dropwise to the round-bottom flask from the addition funnel. Refluxing was continued for 24 hours. After completion of the Quaternization reaction, most of the acetonitrile solvent was removed from the system by rotary evaporation. Finally, the remaining solvent and colored impurities were removed by the recrystallization process. 1M solution of C4 was prepared in deionized water. The recrystallization process was initiated by adding seed C4 crystal into the solution. The solution was kept at room temperature for a week for complete crystallization. A solid crystalline final product was obtained for benzyl dimethylbutyl ammonium bromide (C4). The structure, purity, and characterization of synthesized benzyldimethylbutyllammonium bromide (C4) Quat were determined by ^{1}H-NMR spectra (**Figure A1 in Appendix A**). The purity of the final product was ~ 99.71%.

Synthesis of Benzyldimethylhexylammonium Bromide (C6)

0.100 mole N, N-Dimethylbenzylamine and 100 ml acetonitrile were added into the three-necked round-bottom flask. Then, a condenser, an addition funnel, and a thermometer were attached to the three necks of the flask. Water flow in the condenser was ensured to reflux acetonitrile and amine mixture. Then, the round-bottom flask was heated to the reflux temperature of 82°C by using a heating mantle. After refluxing commenced, the respective 0.100 moles (equimolar) bromohexane was added dropwise to the round-bottom flask from the addition funnel. Refluxing was continued for 24 hours. After completion of the Quaternization reaction, most of the acetonitrile solvent was removed from the system by rotary evaporation. Finally, the remaining solvent and colored impurities were removed by the recrystallization process. 1M solution of C6 was prepared in deionized water. The recrystallization process was initiated by adding seed C6 crystal into the solution. The solution was kept at room temperature for a week for complete crystallization. A solid crystalline final product was obtained for benzyl dimethyl hexylammonium bromide (C6). The structure, purity, and characterization of synthesized benzyl dimethyl hexylammonium bromide (C6) Quat were determined by ^{1}H-NMR spectra (**Figure A2**). The purity of the final product was 99.63%.

Synthesis of Benzyldimethyloctylammonium Bromide (C8)

0.100 mole N, N-Dimethylbenzylamine and 100 ml acetonitrile were added into the three-necked round-bottom flask. Then, a condenser, an addition funnel, and a thermometer were attached to the three necks of the flask. Water flow in the condenser was ensured to reflux acetonitrile and amine mixture. Then, the round-bottom flask was heated to the reflux temperature of 82°C by using a heating mantle. After refluxing commenced, the respective 0.100 moles (equimolar) bromooctane was added dropwise to the round-bottom flask from the addition funnel. Refluxing was continued for 24 hours. After completion of the Quaternization reaction, most of the acetonitrile solvent was removed from the system by rotary evaporation. Finally, the remaining solvent and colored impurities were removed by the recrystallization process. 1M solution of C8 was prepared in deionized water. The recrystallization process was initiated by adding seed C8 crystal into the solution. The solution was kept at room temperature for a week for complete crystallization. A solid crystalline final product was obtained for benzyl dimethyl octylammonium bromide (C8). The structure, purity, and characterization of synthesized benzyl dimethyl octylammonium bromide (C8) Quat were determined by ^{1}H-NMR spectra (**Figure A3**). The purity of the final product was ~ 99.67%.

Synthesis of Benzyl Dimethyldecylammonium Bromide (C10)

0.100 mole N, N-Dimethylbenzylamine and 100 ml acetonitrile were added into the three-necked round-bottom flask. Then, a condenser, an addition funnel, and a thermometer were attached to the three necks of the flask. Water flow in the condenser was ensured to reflux acetonitrile and amine mixture. Then, the round-bottom flask was heated to the reflux temperature of 82°C by using a heating mantle. After refluxing commenced, the respective 0.100 moles (equimolar) bromodecane was added dropwise to the round-bottom flask from the addition funnel. Refluxing was continued for 24 hours. After completion of the Quaternization reaction, most of the acetonitrile solvent was removed from the system by rotary evaporation. The final product was ionic liquid for benzyl dimethyl decylammonium bromide (C10). The structure, purity, and characterization of synthesized benzyl dimethyl decylammonium bromide (C10) Quat were determined by ^{1}H-NMR spectra (**Figure A4**). The purity of the final product was more than ~ 99.27%.

Synthesis of Benzyl Dimethyl Dodecylammonium Bromide (C12)

0.100 mole N, N-Dimethylbenzylamine and 100 ml acetonitrile were added into the three-necked round-bottom flask. Then, a condenser, an addition funnel, and a thermometer were attached to the three necks of the flask. Water flow in the condenser was ensured to reflux acetonitrile and amine mixture. Then, the round-bottom flask was heated to the reflux temperature of 82°C by using a heating mantle. After refluxing commenced, the respective 0.100 moles (equimolar) bromododecane was added dropwise to the round-bottom flask from the addition funnel. Refluxing was continued for 24 hours. After completion of the Quaternization reaction, most of the acetonitrile solvent was removed from the system by rotary evaporation. The final product was ionic liquid for benzyl dimethyl dodecylammonium bromide (C12). The structure, purity, and characterization of synthesized benzyl dimethyl decylammonium bromide (C12) Quat were determined by ^{1}H-NMR spectra (**Figure A5**). The purity of the final product was ~ 99.03%.

Synthesis of Imidazolines from Carboxylic Acids

Five imidazoline-based surfactants from C4 to C12 chain length were synthesized in a similar reaction condition. For example, the synthesis of C12 imidazoline was described here. The reaction was carried out in neat reaction conditions. No solvents were added to the reactants. About 0.04 mol of tridecanoic acid (for C12) was placed in a 100 ml, 3-neck flask, equipped with a thermocouple, addition funnel, and a Dean-Stark trap. At first, tridecanoic acid was heated to about 70°C under N_2 and then about 0.06 mol of diethylenetriamine (DETA) was added dropwise rapidly (tridecanoic acid: DETA was 1:1.5) from the addition funnel. The resulting mixture changed from colorless to yellow color and temperature increased to100 °C. The mixture was then heated to about ~180 °C for 3 hours for imidazolines precursor (amide) formation while allowing water to collect in the Dean-Stark trap. Finally, the resulting mixture was then heated at about ~230 °C for an additional 2 hours imidazolines ring closure and further evolved water was collected. A reaction scheme for the imidazolines synthesis is shown in **Figure 2.3**.

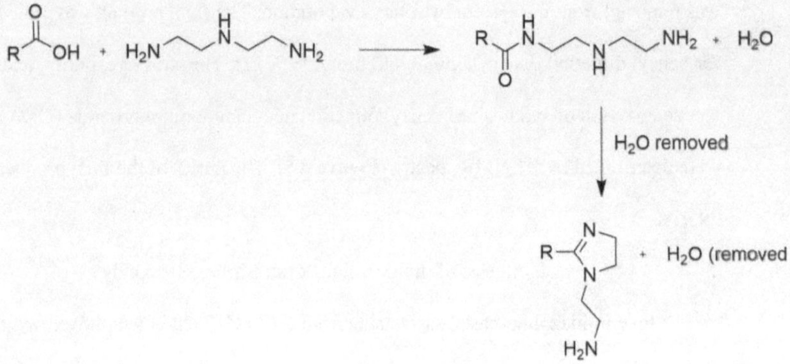

Figure 2.3: Synthesis of imidazolines from fatty acid and DETA.[51-53]

The structure and purity of the final product was characterized using ^{13}C-NMR. The amide (C=O) and imidazoline (C=N) characteristic peaks at ~172 ppm and ~ 167 ppm were integrated for comparison. The ^{13}C-NMR characterization of the final product showed ~90% imidazolines and ~10% amide mixture in **Figure 2.4**.[51, 52, 54]

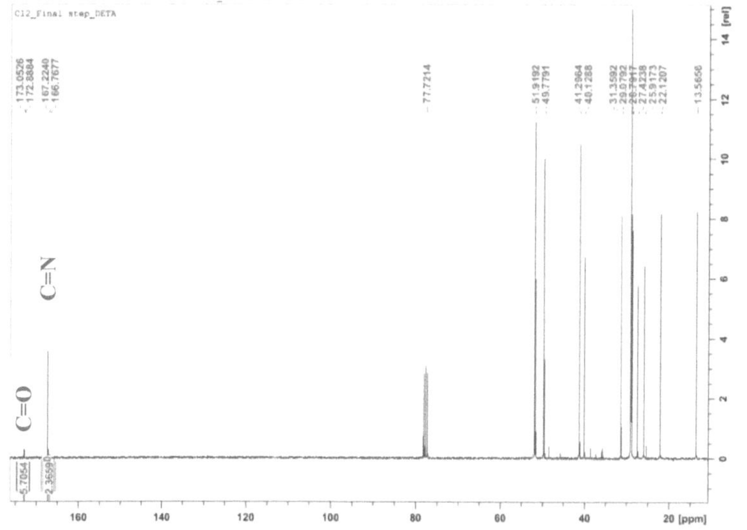

Figure 2.4a: ^{13}C-NMR spectrum of imidazoline (C12).

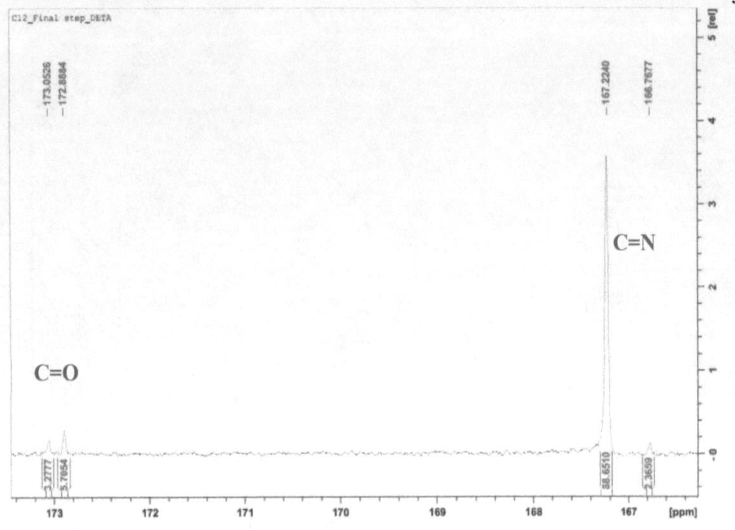

Figure 2.4b: ^{13}C-NMR spectrum of imidazoline (C12)-examined closely.

Synthesis of Imidazolines from Aldehydes (One-pot Synthesis)

Imidazolines corrosion inhibitor was synthesized by the one-pot synthesis method. Nitriles and esters were used as starting materials to avoid some difficulties with toxic cyanide, a high reaction temperature, or acidic conditions are required. One-pot synthesis of imidazolines in tert-butyl methyl ether or dichloromethane (CH_2Cl_2) proceeds in excellent yield. At first condensation of aldehydes and diamines in dichloromethane formed aminals. Next, N-bromosuccinimide (NBS) was added to oxidize aminals to imidazolines in the one-pot operation. The reaction was carried out at the low reaction temperature (0°C - room temperature). A reaction scheme for the imidazolines synthesis is presented below (**Figure 2.5**).[55] 2-Bromoethylamine hydrobromide in acetonitrile was added to the imidazoline ring to get the desired product.[56] The reaction was quenched by the addition of saturated $Na_2S_2O_{5(aq)}$, and 10% $NaOH_{(aq)}$ solutions. The mixture was extracted with CH_2Cl_2. The extracted organic layer was then dried over Na_2SO_4 and evaporated in a vacuum oven.

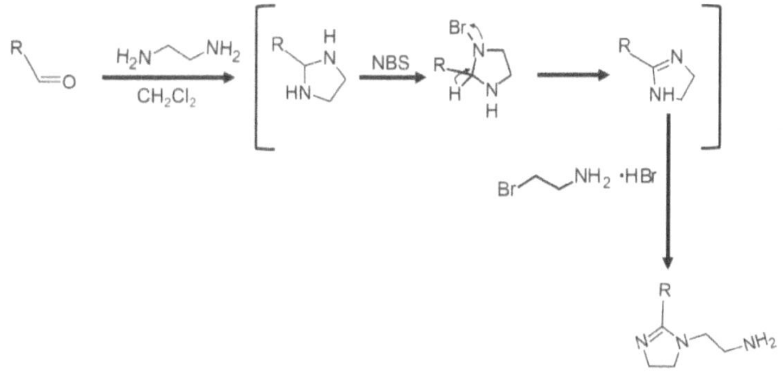

Figure 2.5: One-pot synthesis of imidazolines.[55, 56]

Synthesis of C12 Imidazolines using Aldehydes (One-pot synthesis)

Five imidazolines based surfactants from C4 to C12 chain length were synthesized in a similar reaction condition. For example, the synthesis of C12 imidazoline was described here. Pentanal (1 mmol) was added to the mixture of diamine (1.05 mmol) in 10 ml CH_2Cl_2 and stirred at $0°C$ for 30 minutes under N_2 gas. Next, NBS (1.05 mmol) was added to the mixture and stirred overnight at room temperature. The final product was obtained by 2-Bromoethylamine hydrobromide in acetonitrile solvent.

The mixture was extracted with CH_2Cl_2 and characterized using ^{1}H-NMR and ^{13}C-NMR (**Figure 2.6** and **Figure 2.7**).

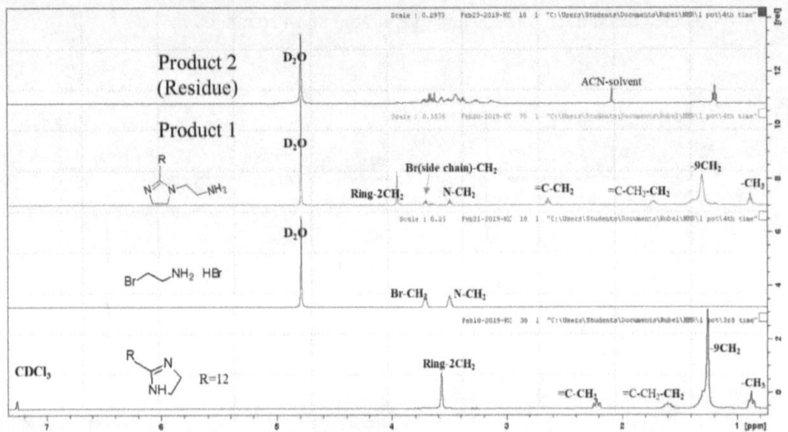

Figure 2.6 ^{1}H-NMR spectra of Imidazoline (C12)

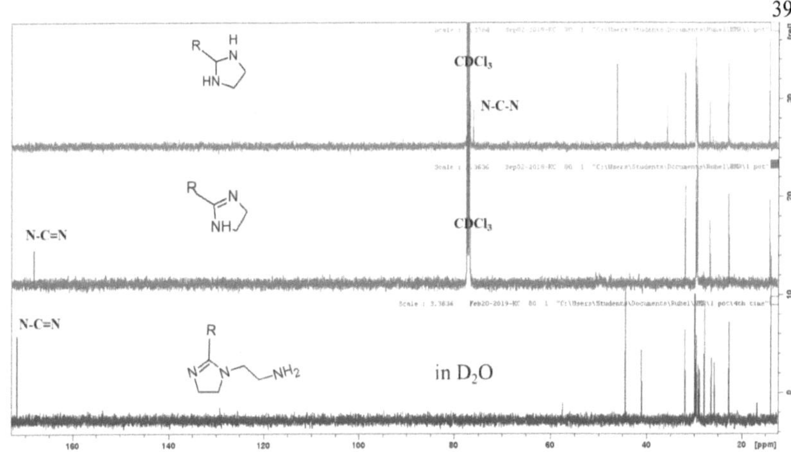

Figure 2.7. ^{13}C-NMR spectra of Imidazoline (C12)

CHAPTER 3: EXPERIMENTAL METHODS

SFG Instrumentation

In Cimatu surface science laboratory, the SFG spectra are acquired using a custom-built broadband SFG scheme using a femtosecond laser.[24, 29] The schematic diagram of the spectroscopic set-up is shown in **Figure 3.1**. The femtosecond laser is a one-box ultrafast amplifier system (Solstice) from Spectra-Physics. The Solstice consists of a regenerative amplifier (RGA), Mai Tai as a seed laser, Empower as a pump laser, stretcher, and compressor. The output of Mai Tai is a tunable wavelength of 795 nm beam with <100 fs pulse duration. After the pulse stretching, the seed laser in RGA is amplified by a pump laser followed by compression, the output generates a 795 nm wavelength of light with 100 fs duration pulses, 1kHz repetition rate, and an average ~4 W output power. Then, the amplified Solstice beam passed through a 50:50 beam splitter. One-half of the beam is used to pump an automated optical parametric amplification system (TOPAS-C, Light Conversion) and is coupled collinearly to difference frequency generation (DFG) crystal (AgGaSe2) to generate a tunable broad-bandwidth mid-infrared beam. The mid-infrared beam is tunable from 4000 to 1000 cm^{-1} (2.5-12 μm) with a full-width half-maximum (fwhm) of 200-250 cm^{-1}. The remaining half of the amplified 795 nm beam is passed through a Fabry-Perot etalon (SLS optics, FSR 30.931 nm, Re 95.69 at 800 nm) that generates a narrow-bandwidth picosecond pulse and has a fwhm of ~7 cm^{-1}. After that, both beams (795 nm and mid-infrared) are aligned and focused to the sample surface at 50° and 60° respectively to generate the broadband SFG beam at 52° when the IR pulses are centered at 3375 nm, as shown in **Figure 3.1**.

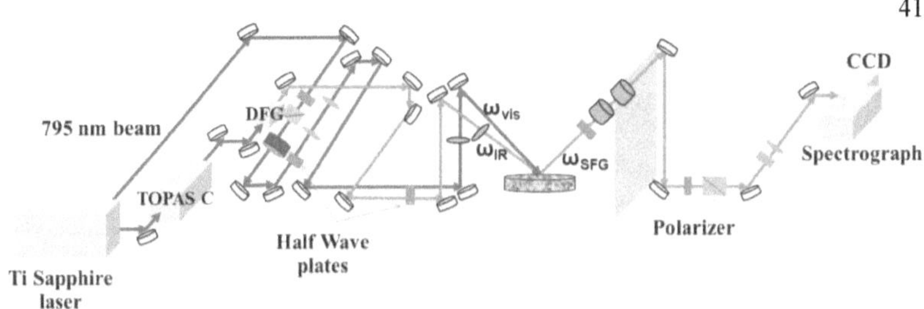

Figure 3.1: SFG experimental set-up

Sample Preparation

B.1: Preparation of eicosanoic acid monolayer on water (air-liquid interface)

A self-assembled eicosanoic acid monolayer on deionized water was used as a reference sample for the SFG signal optimization and SFG signal intensity normalization at the air-liquid interface.

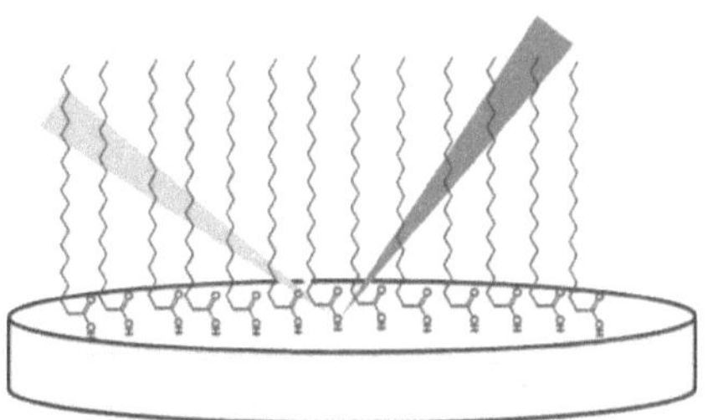

Figure 3.2: Eicosanoic acid monolayer on water at air-liquid interface

Eicosanoic acid (EA) more than 99% purity; was purchased from Sigma Aldrich (St. Louis MO) and used as received. 2.00 mg of EA was dissolved in 1.00 ml of chloroform.

25.00 µl of this solution was added to a petri dish containing 60.00 ml of deionized water to form a self-assembled monolayer. The SFG spectrum was collected using the ssp polarization combination.

Preparation of Octadecanethiol (ODT) Monolayer on Gold (liquid-solid interface)

The Octadecanethiol monolayer on gold thin film was used as a reference sample for the liquid-metal interfaces. 1 mM of ODT solution in 200 proof ethanol (purchased from Fisher Scientific) was prepared in a 40 ml vial. The cleaned pristine100 nm gold thin film on a silicon wafer (purchased from Sigma Aldrich) was immersed in the ODT solution for 18 hours for full monolayer formation. Then, the ODT monolayer on gold was cleaned with ethanol to remove residual ODT solution on the gold substrate.

Metal Surface Preparation

Mild Steel Preparation

The commercial-grade mild steel similar to oil and gas transportation pipelines were taken for SFG characterization. Small pieces (~1.5 cm) of square-shaped were cut using Dremel tools. The mild steel piece was polished with sandpaper and 1200 grit carbon paper. Then, the mild steel was sequentially polished using 6 microns, 3 microns, 0.5 microns, and point 0.1-micron diamond pastes. Finally, the mild steel was further polished with 0.05-micron aluminum slurry. The polished mild steel was sonicated with acetone, thoroughly rinsed with deionized water and ethanol, dried with nitrogen gas, and placed in an SFG cell.

Stainless Steel Preparation

The pre-cut stainless steel was sequentially polished using 6 microns, 3 microns, 0.5 microns, and point 0.1-micron diamond pastes. Finally, the stainless steel was further polished with 0.05-micron aluminum slurry. The polished stainless steel was sonicated with acetone, thoroughly rinsed with deionized water and ethanol, dried with nitrogen gas, and placed in an SFG cell.

Iron (Fe) Surface Preparation

The pre-cut (1-inch x 1inch) Fe metal (99.95%) was used for SFG experiments. The Fe piece was sequentially polished using 6 microns, 3 microns, 0.5 microns, and point 0.1-micron diamond pastes. Finally, the metal was further polished down to 0.05 micron with aluminum slurry. The polished Fe was sonicated with acetone, thoroughly rinsed with deionized water and ethanol, dried with nitrogen gas, and placed in an SFG cell. Dremel tools were used instead of scissors to avoid bending towards the edges.

SFG Sample Cell for Liquid-metal Interfaces

A sample cell in Teflon was used for the liquid-metal interface experiment in **Figure 3.2.**[57]

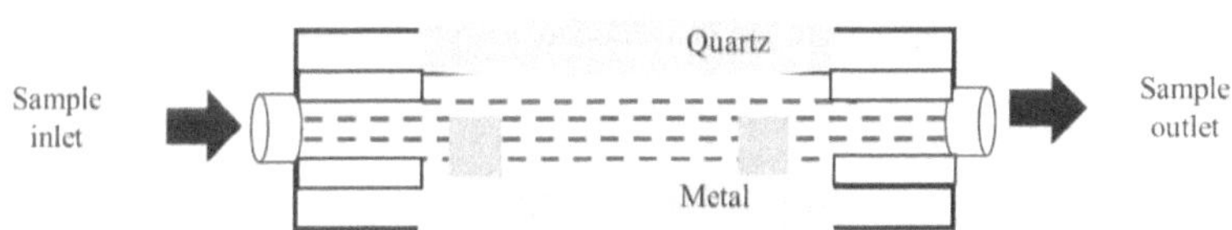

Figure 3.3: Cross-sectional representation of SFG sample cell for liquid-solid interface

The cell contains an inlet, an outlet, a fused silica window that helps seal the cell, an adjustable Teflon spacer, and a sample holder with an adjustable height to maintain a constant volume of aqueous solution between the window and the metal surface. The freshly prepared Quat solution (~5 ml) in D_2O was injected into the sample cell. The solution is equilibrated for an hour inside the sample cell before collecting the SFG data. The overlap of both visible and IR beams was optimized at the liquid-gold metal interface by using a reference beam (HeNe) and SFG spectra were collected *in-situ*.

SFG Data Analysis

Data Collection

For the SFG experiment, a gold substrate is used to optimize the overlap between the visible and IR beams. Before the SFG experiment, the gold substrate is treated with ethanol, dried with nitrogen gas, and plasma cleaned oxygen gas. The signal from gold thin film provides perfect optimization of both beams at the sample stage and IR beam profiles at each IR center. All the spectra are collected for three trials with a background at ssp and ppp polarization combinations. Each trial is collected for 3 seconds acquisition time for 180 accumulations giving a total integrated time of 9 minutes.

Processing of Raw Data

The experimental raw spectra are collected using a commercial Andor SOLIS data acquisition software and converted to .ascii files for further processing. The converted .ascii files are exported to OriginPro 9.1 software. The spectra are subtracted from the background, the x-axis is corrected using polystyrene, smoothed at every 10 points using Savitzky- Golay function, and corrected for polarization efficiency of optics. The average SFG intensity of three trials is reported with the error bars. The x-axis represents the IR wavenumber (cm^{-1}) and the y-axis represents the SFG Intensity (a.u.).

Fitting

The SFG spectra are fitted with the Lorentzian line shape equation using Mathematica software. The simplified version of the fitting equation that accounts for the CH and OH vibrational modes is presented below.

$$I_{SFG}\,(\omega + \omega_{vis}) \propto exp\left[-\frac{(\omega-\omega_{IR}^{L})^2}{2(\delta\omega_L)^2}\right] \times \left|\Sigma_q \frac{A_q}{\omega_{IR}-\omega_q+i\Gamma_q} + A_{NR}e^{i\rho}\right|^2 \qquad (3.1)$$

Where $\delta\omega_L$ is the spectral width of the broadband pulse, A_q is the amplitude of the vibrational mode, ω_q is a vibrational frequency, ω_{IR} is the incident IR frequency, Γ_q is peak width, A_{NR}, and ρ is the amplitude and phase of the nonresonant contribution. The nonresonant contribution from the bulk and the Gaussian equation are both included in the fitting equation to account for the mid-IR beam's broadband pulse width.

Surface Tension Measurement

The solubility, surface coverage, and aggregation behavior of surfactants can be predicted from The critical micelle concentration (CMC) of surfactants. The CMC values of Quat solutions were determined by measuring surface tension values. The CSC scientific Du Noüy tensiometer was used for the surface tension experiments. The glassware used for these experiments were acid cleaned overnight, then rinsed and boiled with deionized water. The platinum ring was cleaned by rinsing it in isopropanol and then heat in the oxidizing portion of a gas flame. The ring was cleaned between each measurement. The Quat solutions were prepared at different concentrations in deionized water (~ 45 ml) in a 60 ml petri dish. The ring was immersed in the solution containing a petri dish and the torsion of the ring was increased until the film break. The subsequent surface tension value was measured from the scale. The surface tension values were measured as a function of concentrations. All the surface tension values were collected for three trials and corrected with the correction factor given for the instrument.

pH Measurement

The pH of Quat solutions was measured using a Datalogger connected to a pH sensor (Xplorer GLX) purchased from PASCO. The pH sensor was calibrated with pH calibration standards (pH 4.00 and pH 10.0). Quat solutions were prepared in deionized water. Then, the pH of Quat solutions was measured at room temperature with the same setup. The data were collected for three trials and the average values were reported in Appendix B, **Table B.4.**[49]

CHAPTER 4: ROLE OF THE CATIONIC HEADGROUP TO CONFORMATIONAL CHANGES UNDERGONE BY SHORTER ALKYL CHAIN SURFACTANT AND WATER MOLECULES AT THE AIR-LIQUID INTERFACE

Introduction

Surfactants are useful in applications such as corrosion inhibition, soaps, paints, adhesives, inks, insecticides, antibacterial, antifungal, and many other applications.[58-60] Depending on the application, surfactant molecules are being constantly modified to serve a specific purpose under specific condition. The combined use of surface tension and SFG shall help us to understand how alkyl chain length and ionic strength affect the self-assembly, surface coverage, and CMC of surfactants. These fundamental studies of surfactants at the air-water interface will provide novel insights to design new surfactants for a specific field of application.

Surfactants are widely employed as corrosion inhibitors to protect metal surfaces of pipelines in the oil-and-gas industry due to their low toxicity, high solubility, high hydrocarbon content, and ability to adsorb onto metals to retard corrosion.[6-8, 14, 20, 30, 61] These surfactants can react with the impurities of the metal surface or restrict the rate of the anodic or cathodic processes by simply blocking the active sites or pits present on metal surfaces.[8] The most commonly used inhibitors are from groups of amines, amides, Quaternary ammonium salts, imidazolines, aromatic aldehydes, acetylenic alcohols, condensation products of carbonyls, and alkyl trimethylammonium-based surfactants.[7, 16, 32-34] These compounds are amphiphilic by nature, which contain a hydrophilic headgroup and a hydrophobic chain group. As the headgroup adsorbs on the metal surface forming a self-assembled monolayer, the alkyl chain group arranges itself towards the bulk solution and provides a barrier for corrosive substances like water and dissolved gases (CO_2 and

SO$_2$) using electrochemical impedance spectroscopy.[14, 16] Thus, these surfactants become good candidates for corrosion inhibition.[15]

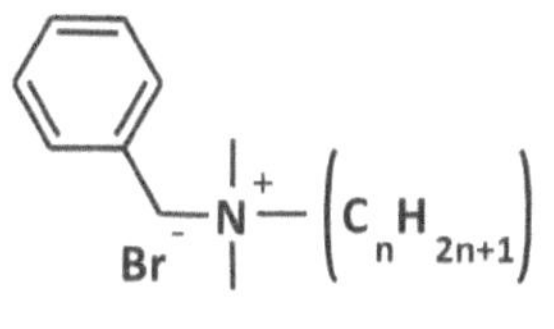

Figure 4.1. Quaternary ammonium bromides (Quats), where n = 4, 6, 8, 10, and 12.

Both head and alkyl chain groups play an important role in conformational changes of surfactants at air-liquid and solid-liquid interfaces.[39, 43] In water-soluble surfactants, the chain groups project to the air and the headgroup points to the interfacial water at the air-water interface. The presence of charged surfactants at the air-water interface shows an overall enhancement of hydroxy (OH) peaks of interfacial water molecules.[62] Charged surfactants can create a large electrostatic field in the double-layer region. In this scenario, interfacial water molecules orient themselves with the hydrogen of OH bands pointing down for cationic surfactants, whereas the hydrogen of OH bands for anionic surfactants is pointing up.[63] The previous works on conformational changes of surfactants and water molecules by varying the headgroup, alkyl chain group, concentration, and ionic strength using sum frequency generation (SFG) spectroscopy have not reported the combined effect of above-mentioned factors.[43, 64-69] However, the chain length dependence of highly soluble surfactants with fewer carbons (C4 – C12) at various salt concentrations in water has not yet been extensively studied at the air-liquid interface. Most of these surfactants do not reach CMC even at a high concentration which is the key factor for maximum surface coverage. The interfacial behavior of salt-water-surfactant at different chain lengths studied by surface-specific sum frequency generation spectroscopy (SFG) will be of great interest. Surfactants in water create an electrostatic field that greatly orients water molecules at the interfacial and double-layer regions; also, the presence of salt disturbs the hydrogen-bonding network of water.[70] The combined influences of the chain length and ionic strength in water should provide new insights into adsorption, self-assembly, and interfacial process like corrosion reaction.

Independent studies on chain length and ionic strength in both air-solid and air-liquid interfaces state that as the number of carbon atoms increases, a decrease in the number of gauche defects within the alkyl chains is expected.[67, 68] Another study, performed by Richmond and co-workers, reported a decreasing number in gauche conformations with decreasing chain length at the liquid-liquid interface.[69] In some other cases, such as studies with neat ionic liquids, it has been observed that the relative number of gauche defects increases with increasing chain length.[64, 71] On the other hand, the ionic strength dependence shows that the intensity of the OH stretch band positioned at $\sim$ 3200 cm^{-1} of the interfacial water at a charged surface decreases with increasing salt concentration due to the screening effect.[50, 62, 72-74] Thus, very few studies were performed that mainly focused on varying the chain length from C4- C12 in the absence and presence of inorganic salt at the air-water interface. Also, studies on increasing chain length are somehow inconclusive when the number of gauche defects is concerned at different interfaces. Thus, it becomes noteworthy to learn how the conformation of shorter-chained surfactants and water molecules are arranged at the air-liquid interface in relation to their ability to form a well-ordered monolayer, for the same reason that shorter chain alkyl surfactants are more soluble in water and becomes an ideal inhibitor because of its solubility properties. The insight it provides is important especially for applications such as corrosion inhibition.

In this study, we investigated conformational changes and monitored spectral profiles of Quaternary-based ammonium surfactants with varying chain lengths from C4 – C12 in aqueous solution, and varying salt (NaCl) concentrations at the air-water interface in CH and OH regions by SFG spectroscopy. SFG spectra were acquired from 2800 cm^{-1} to 3700 cm^{-1} with ssp and ppp polarization combinations. Each SFG spectrum was fitted to extract amplitude values in order to estimate gauche defects between the ratio of the methyl symmetric stretch (CH_3 SS) and the methylene symmetric stretch (CH_2 SS) and to determine the average tilt angle value of the CH_3 group for the C4 Quat molecule. The fitted values were also useful in plotting the OH stretch as a function of chain length as well as salt concentration. Surface tension measurements were also performed for most of the aqueous solutions of the surfactant to calculate the surface coverage and find its correlation with SFG results.

SFG Theory

Sum frequency generation (SFG) spectroscopy is a surface-specific vibrational technique that obtains the vibrational spectra of the interfacial molecules.[24] The spectral analysis provides information about identification, conformation, and dynamics of interfacial molecules.[25-27, 75] SFG spectroscopy provides molecular surface specificity and sub-monolayer sensitivity. It involves a nonlinear optical process in which its intrinsic symmetry requirement dictates that SFG can only arise from an interface, noncentrosymmetric bulk crystals, or bulk chiral molecules.[75] The centrosymmetry of the bulk can be explained by the isotropic distribution of molecules, and thus this results in no SF signal arising from the bulk. However, this symmetry breaks at the interface, thus creating an asymmetric environment for these interfacial molecules and becomes SF active.

The SFG signal is generated at the sample surface from the temporal and spatial overlap of the two pulsed laser beams. One laser beam is the visible beam with the frequency; ω_{VIS} and the other beam is the mid-infrared beam with tunable frequency; ω_{IR} and the sum of the two frequencies results in the generated output beam of the SFG process (ω_{SF}).

$$\omega_{SF} = \omega_{VIS} + \omega_{IR} \qquad (4.1)$$

SFG intensity is directly proportional to the square of the effective second-order nonlinear susceptibility, with the incident fixed 795 nm and mid-IR beams.

$$I_{(\omega_{SF})} \propto \left[\sum_i \chi_{eff}^{(2)} \sum_{j,k} E_j E_k \right]^2 \qquad (4.2)$$

Herein, $I_{(\omega_{SF})}$ is the SFG intensity and $\chi^{(2)}$ is the effective second-order nonlinear susceptibility, $\chi^{(2)}$. $\chi_{eff}^{(2)}$ is a third rank tensor which can be further defined by using Fresnel factors and the macroscopic nonlinear susceptibilities, $\chi_{ijk}^{(2)}$, at the interface, as shown in equation 4.3.

$$\chi_{eff,ijk}^{(2)} = \chi_{ijk}^{(2)} [L_i e_i][L_j e_j][L_k e_k] \qquad (4.3)$$

Where i, j, and k coordinates represent x, y, z-axes, respectively. L represents the Fresnel factor while e represents the unit optical field vector. The macroscopic susceptibility has

27 components, but due to symmetry constraints obtained at the surface, some of the components become almost negligible. Therefore, there are only 7 components left, which correspond to four possible polarization combinations; ssp, ppp, pss, and sps. SFG spectra were collected at ssp and ppp polarization combinations only. The beam polarizations s and p refer to the incident beam perpendicular and parallel to the plane of incidence, respectively.[50] The three beams are listed in the order of decreasing energy; ssp refers to the SFG beam is (s) polarized, the visible beam is (s) polarized, and the mid-IR beam is (p) polarized.

$\chi_{\text{eff}}^{(2)}$ is composed of the resonant susceptibility, $\chi_R^{(2)}$, and the nonresonant susceptibility, $\chi_{\text{NR}}^{(2)}$.

$$\chi_{eff}^{(2)} = \chi_R^{(2)} + \chi_{NR}^{(2)} \tag{4.4}$$

The nonresonant term, $\chi_{\text{NR}}^{(2)}$, depends on the surface plasmon resonance of the substrate and is also found to be invariant with IR frequency.[25] $\chi_{\text{NR}}^{(2)}$ is almost negligible for liquid and dielectric surfaces.

$$\chi_R^{(2)} = \sum_q \frac{N\langle \beta^{(2)} \rangle}{(\omega_{IR} - \omega_q + i\Gamma_q)} \tag{4.5}$$

The resonant term, $\chi_R^{(2)}$, as shown in equation 5, is due to vibrational transitions of the interfacial molecules. N is the number density of vibrational modes, $\beta^{(2)}$ is the orientational averaged hyperpolarizability, and Γ_q is the damping constant of qth vibrational mode, respectively. ω_{IR}, ω_q are the IR frequency and the frequency of the qth vibrational mode, respectively.[24]

Experimental Details

Synthesis and Purification of Alkylbenzyldimethylammonium Bromide (Quats)

Alkylbenzyldimethylammonium bromides containing five different alkyl chain lengths butyl (C4), hexyl (C6), octyl (C8), decyl (C10), and dodecyl (C12) were synthesized using established methods.[47, 50] 0.100 mol of N, N-dimethylbenzylamine (Fisher Scientific, New Hampshire, US) and 100 ml of acetonitrile (Fisher Scientific, New Hampshire, US) were added into the three-necked round-bottom flask. After that, the solution in the round-bottom flask was heated to a reflux temperature of 82°C. At ~ 82 °C, 0.100 mole (equimolar) of bromoalkane (Fisher Scientific, New Hampshire, US) was added dropwise to the solution. Refluxing was continued for 24 hours. The solvent was recovered from the reaction mixture by rotary evaporation. Finally, the trace amount of solvent and colored impurities were removed by recrystallization process with deionized water. A solid crystalline Quat was obtained for C4, C6, and C8 and viscous liquid for C10 and C12. The structure and the purity of synthesized Quats were confirmed by ^{1}H-NMR spectra with their corresponding purity values available in Table B1 of the appendix.[50]

Instrumentation

The SFG set-up utilized Solstice broadband femtosecond Titanium-Sapphire laser from Spectra-Physics (Santa Clara, CA), which generated the fundamental beam at 795 nm. The detailed experimental set-up was described in the earlier publications of the group.[24, 29] The 795 nm visible output beam was sent to a 50:50 beam splitter. The 50% reflected beam passed through a Fabry–Pérot etalon to generate the narrow bandwidth visible beam. The 50% transmitted beam was sent to an optical parametric amplification system (TOPAS-C) and difference frequency generation (DFG) crystal to generate the broadband mid-infrared beam. Both visible and IR beams were temporally and spatially overlapped at the sample surface with incident angles of 50° and 60°, respectively, which generated the SFG beam at 52° from the surface normal. Then the SFG beam was collected through a spectrograph and a charge-coupled device camera.

Sample Preparation

Eicosanoic acid (EA) with ≥99% was purchased from Sigma Aldrich (St. Louis MO) and used as received. 2 mg of EA was dissolved in 1 ml of chloroform. 25 µl of this solution was used to form a monolayer in a petri dish containing 60 ml of deionized water. A self-assembled EA monolayer on the water was used as a reference sample for the SFG signal optimization at the air-liquid interface. The SFG spectrum was collected using the ssp polarization combination.[24, 29] 8.0 mM aqueous solutions (60 mL) in deionized H_2O (18 MOhm cm) were prepared for the five Quats compounds (C4, C6, C8, C10, and C12) with 0%, 1% and 10% (w/w) of sodium chloride (NaCl). Additional solutions with 0.5% and 5% sodium chloride were prepared for the C8 Quat compound. Freshly prepared samples were sonicated for ~20 minutes. The EA solution in a glass Petri dish was replaced by an equal volume of Quat aqueous solution in a similar glass Petri dish and equilibrated for 30 minutes prior to collecting SFG spectral data.

Data Acquisition

The gold and polystyrene films were used to optimize the overlap of the two beams and to calibrate the x-axis for the IR frequency, respectively. Gold spectra were collected at ten IR centers for every 100 cm^{-1} from 2800 cm^{-1} to 3700 cm^{-1} with one-second acquisition time. All spectra were background-corrected and stitched together.

SFG spectra of the five Quaternary ammonium bromide samples were acquired at ten IR centers from 2800 cm^{-1} to 3700 cm^{-1} at every 100 cm^{-1} to probe the entire CH and OH regions using ssp and ppp polarization combinations. Three trials and background were collected for each run. Then, each trial was collected for 9 minutes (3-sec acquisition time for 180 accumulations). The energies of the visible beam and IR beam were measured to be ~26 μJ and ~8 μJ, respectively, at the sample stage.

Fitting Equation

The following Lorentzian line shape equation is considered to analyze each polarization combination and also accounts for the line broadening.

$$I_{SFG} \propto \left|\chi^{(2)}\right|^2 \propto \left|\sum_q \frac{N\langle\beta^{(2)}\rangle}{(\omega_{IR}-\omega_q+i\Gamma_q)} + \left|\chi_{NR}^{(2)}\right|e^{i\rho}\right|^2 \qquad (4.6)$$

Where ρ is the phase of the nonresonant response. The nonresonant contribution from the bulk and the Gaussian equation are both included in the fitting equation to account for the mid-infrared beam's broadband pulse width. The simplified version of the fitting equation that accounts for the CH and OH vibrational modes is presented below.

$$I_{SFG}(\omega + \omega_{vis}) \propto exp\left[-\frac{(\omega-\omega_{IR}^L)^2}{2(\delta\omega_L)^2}\right] \times \left|\sum_q \frac{A_q}{\omega_{IR}-\omega_q+i\Gamma_q} + A_{NR}e^{i\rho}\right|^2 \qquad (4.7)$$

Where, $\left|\sum_q \dfrac{A_q}{\omega_{IR}-\omega_q+i\Gamma_q}\right|^2 = \left|\sum_{CH}\dfrac{A_{CH}}{\omega_{IR}-\omega_{CH}+i\Gamma_q} + \sum_{OH}\dfrac{A_{OH}}{\omega_{IR}-\omega_{OH}+i\Gamma_q}\right|^2$

The Gaussian curve equation is defined with the spectral width $\delta\omega_L$ centered at ω_{IR}^L. The spectral width $\delta\omega_L$ centered at ω_{IR}^L is included in the Gaussian function. The amplitude factors, A_q and A_{NR}, are proportional to the molecular hyperpolarizabilities as shown above in equation 5.7. The amplitudes, A_{CH} and A_{OH}, and frequency positions, ω_{CH} and ω_{OH}, result from CH and OH vibrational modes.

Results and Discussion

As discussed above, the Quaternary ammonium bromides (Quats) surfactants of interest for this study are Quats with 4, 6, 8, 10, and 12 carbons for the hydrophobic portion of the amphiphilic compound. Since these Quats are useful in preventing internal corrosion of oil-and-gas transportation pipelines, an insight towards its adsorption, self-assembly, and behavior at an interface is beneficial towards understanding the role of these shorter chain length surfactants in an aqueous phase.

Peak Assignments

8.0 mM solutions of five Quats in H_2O (~60 mL) were characterized at the air-liquid (AL) interface by SFG spectroscopy. **Figure 4.2** shows the SFG spectra taken at the ssp and ppp polarization combinations. The vibrational modes arise from the terminal methyl and methylene groups of the chain (C4-C12), the two ammonium methyl groups, and benzyl unit of the headgroup **(Figure 1),** and OH stretching modes of hydrogen-bonded water molecules. The Quat and water molecules were probed in CH and OH regions. The peak assignments were described elsewhere.[50] The peaks of the ssp spectrum (**Figure 4.2a**) are assigned from left to right, first two peaks can be assigned to the methylene symmetric stretch (CH_2 SS) at ~2851 cm^{-1} [76, 77] and terminal methyl symmetric stretch (CH_3 SS) at ~2876 cm^{-1}.[76-79] The next two peaks can then be assigned to the methylene asymmetric stretch (CH_2 AS) at ~2915 cm^{-1} and methyl Fermi resonance (CH_3 FR) at ~2936 cm^{-1}. CH_3 FR at ~2936 cm^{-1} is the Fermi resonance of stretching and bending overtones (CH_3 SS and the overtone of the methyl bending mode).[27, 76, 77] The fundamental mode (methyl symmetric stretch) is split by Fermi resonance with an overtone of a methyl symmetric bending mode. This interaction causes a low frequency peak (CH₃ SS at ~2876 cm⁻¹) and a high frequency peak (CH₃ FR at ~2936 cm⁻¹).[25] Fermi resonance is the change of intensities and energies of vibrational modes observed in the SFG spectrum as a result of quantum mechanical wavefunction mixing.[80] Fermi resonance mostly occurs between the fundamental and overtone of the same vibrational group. Due to the Fermi resonance, the vibrational mode with high energy shifts to higher energy and the vibrational mode with low energy shifts to lower energy. Also, the weaker vibrational mode gains intensity from the stronger vibrational mode.

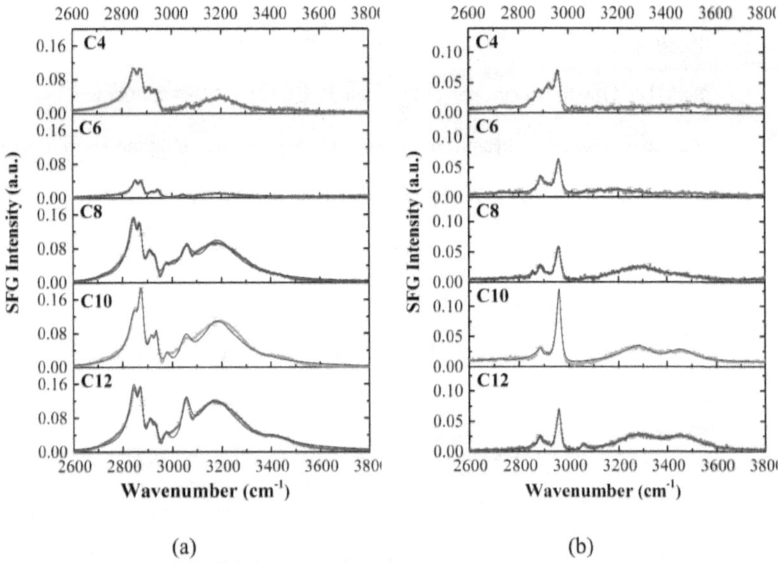

(a)   (b)

Figure 4.2. Fitted SFG spectra of 8mM C4, C6, C8, C10, and C12 Quat compounds in water (without NaCl) at (a) ssp and (b) ppp polarization combinations.

The peaks observed at ~2978 cm[-1] and ~3058 cm[-1] were assigned to ammonium methyl symmetric stretch (N-CH$_3$ SS) and ammonium methyl asymmetric stretch (N-CH$_3$ AS) or the combination of aromatic CH stretch and N-CH$_3$ AS, respectively.[81, 82] The peaks at ~3182 cm[-1] and ~3384 cm[-1] are assigned as OH stretching modes for tetrahedrally and asymmetrically hydrogen-bonded water molecular arrangements, respectively.[83-86]

In the ppp polarization combination in **Figure 2b**, the peaks positioned at ~2857 cm[-1], ~2889 cm[-1], ~2915 cm[-1], and ~2961 cm[-1] were assigned to methylene Fermi resonance (CH$_2$ FR),[76] terminal methyl symmetric stretch (CH$_3$ SS), methylene asymmetric stretch (CH$_2$ AS), and methyl asymmetric stretch (CH$_3$ AS), respectively.[77, 87] The CH$_3$ SS in the ppp spectrum is shifted by ~13 cm[-1] compared to ssp spectrum due to Fermi interactions with the overtone of the CH$_3$ bending vibrations, causing unequal SFG intensity redistribution.[88-90] A summary of the peak assignments is available in **Table 4.1**.

Table 4.1. Peak assignments from the fitted SFG spectra of Quats at ssp and ppp polarization combinations.

Peak assignments	ssp	ppp
	Wavenumber (cm^{-1})	Wavenumber (cm^{-1})
CH_2 SS	~2851	-
CH_2 FR	-	~2857
CH_3 SS	~2876	~2889
CH_2 AS	~2915	~2915
CH_3 AS	-	~2961
CH_3 FR	~2936	-
N-CH_3 SS	~2978	-
Aromatic CH stretch (υ_{20b} mode)	-	~3022
N-CH_3 AS /Aromatic CH stretch (υ_2 mode)	~3058	~3066
OH SS (ice-like)	~3182	~3268
OH SS (liquid-like)	~3384	~3451

The antisymmetric υ_{20b} mode of the phenyl group was observed at ~3022 cm^{-1}.[91] The peak positioned at ~3066 cm^{-1} can be either assigned to ammonium methyl asymmetric stretch (N-CH_3 AS)[92] or the symmetric υ_2 resonance of the phenyl group[91] or the combination of the two modes. Both the OH stretching modes of water were also observed in the ppp polarization combination.[83, 84, 86]

Is the conformation of the alkyl chain dependent on the length of Quat surfactants from C4-C12?

Five 8 mM aqueous solutions of Quat compounds with an alkyl chain length varying from C4, C6, C8, and C10 to C12 were prepared for the spectral data acquisition. To test the effects of varying the chain length, we are first analyzing the SFG spectra of the 8 mM Quats in water for any change in the alkyl chain conformation. For well-ordered self-assembled monolayers, the alkyl chain has an all-trans conformation and their CH_2 groups lie in a centrosymmetric environment and thus becoming SF inactive.[25] The SF spectral profile of a well-ordered monolayer mainly contains the terminal CH_3 group vibrational modes whereas, for a not so well-ordered monolayer, the SF spectra comprise of vibrational contributions from CH_3 and CH_2 groups. This conformational disorder in the alkyl chain can be referred to as a gauche defect.[25] Previous studies reported that the extent of gauche defects in the structure of a surfactant monolayer increases with increasing chain length and a further increase in chain length results decrease in gauche defects.[93, 94] In these studies, the enhancement in the CH_2 SS intensity was attributed to an increase in the conformational disorder. Another study reported that the ratio between CH_3 SS to CH_2 SS is inversely proportional to the degree of packing, which they found to be dependent on the chemical nature of the head group.[43] Therefore, to obtain a quantitative insight into the conformation of the Quat compounds at the air-liquid interface, the intensity ratio between CH_2 SS and CH_3 SS was calculated to measure the extent of relative disorder or gauche defects in the alkyl chain conformation.[69] The peak intensities of CH_2 SS and CH_3 SS modes were obtained from

fitted ssp spectra in **Figure 4.2**. A plot of CH_2 SS / CH_3 SS intensity ratios as a function

of the number of carbons in the alkyl chain is shown in **Figure 4.3a**. Intensity ratios

ranging from 0.46 ± 0.16 to 1.5 ± 0.8 indicate that gauche defects were present in all

five Quats solutions at the air-liquid interface. The graph suggests that the formation of

gauche defects has a small dependence on increasing the chain length from C4 to C12. It

is reported in the literature that any chain length from C16 - C22 forms a well-ordered

monolayer.[61] Thus, SFG spectra from these ordered monolayers have minimal gauche

defects.[95] Therefore, the gauche defects obtained from C4 - C12 Quats, which do not

follow a particular trend, as a function of chain length can be due to factors such as the

selected concentration to prepare Quats aqueous solutions and equilibration time.

The critical micelle concentration (CMC) plays an important role in how these

surfactants adsorb and self-assemble at the air-liquid interface. If CMC is not reached, a

full monolayer or coverage will be difficult to confirm. Thus the selection of the bulk

concentration of these interfacial molecules can also be affected by the CMC level. Thus,

if the selected concentration, at 8 mM, is below the CMC value for C4, C6, C8, C10, and

C12, then we expect that the surfactants do not form a full monolayer at the air-liquid

interface. Thus, this reflects the observation of gauche defects without any evident trend.

However, although the CMC values for C4, C6, and C8 are difficult to determine.[61] The

surface tension values of the Quat solutions without salt were obtained using the Du

Noüy ring method and plotted against concentration.[96] Five surface tension plots are

provided in **Figure B.3** in the Supporting Information. **Figures B.3 (a-c)**, plots for C4,

C6, and C8 Quat aqueous solutions did not show more defined transitions to determine their CMC values even though higher concentrations were reached for C4, C6, and C8.

As reported by Nguyen and group, less sharp transitions observed for C4, C6, and C8

solutions can be explained with the assumption that the surface excess/surface coverage

of the topmost monolayer along with the formation of a sub-monolayer is unaffected.

However, the surface tension value is consequently reduced due to the change in the

conformation of the confined interfacial water molecules.[97] As the chain length was

increased, the CMC values for both C10 and C12 are estimated to be ~5 mM and ~0.8

mM, respectively (**Figure B.3 (d)** and **Figure B.3 (e)**). The CMC values show that at

C4, C6, and C8, the spectral profiles collected at 8 mM are all below their undefined

CMC values. On the other hand, since the spectra acquired for both 8 mM C10 and C12

Quat molecules are above their estimated CMC values, the liquid surface is assumed to

be completely covered with Quat molecules. However, as shown in **Figure 4.3(a)**, the

CMC may not have a direct effect towards the number of gauche defects. The number of

gauche defects increases with increasing alkyl chain length. Therefore, the formation of a

well-ordered monolayer in relation to critical micelle concentration does not correlate

with the increased in the number of gauche defects.

In addition, surface excess values were also determined from these surface tension

plots to obtain a better idea of the monolayer coverage of these surfactants at the air-

liquid interface. Therefore, to determine the surface excess concentrations of these

surfactants, the Gibbs adsorption equation (equation 4.8) can be applied.[98-100]

$$\Gamma = -\frac{a}{nRT}\frac{\partial \gamma}{\partial a} = -\frac{c}{nRT}\frac{\Delta \gamma}{\Delta c} \quad (4.8)$$

Where Γ is the surface excess concentration of solutes, a is the activity which can be

replaced with c; the molar concentration of solutes, γ is the surface tension and $n = 2$ for ionic substances. $\Delta\gamma = \gamma_{surfactant\ solution} - \gamma_{pure\ water}$.[100] **Table B.32** available in the Supporting Information lists the surface excess values for each concentration of every chain length. A comparison plot between the surface excess values calculated from the 8 mM concentration of C4 – C12 surfactants and intensity ratios (CH₂ SS/ CH₃ SS) is presented in **Figure B.32**. The surface coverage or surface excess increases with increasing chain length at 8 mM concentration and starts to plateau at C10. The increasing number of surface coverage should have a relatively smaller number of gauche defects because of better packing between the alkyl chains. However, that is not the case for the observation obtained for C4, C6, and C8 Quat solutions. The surface excess/coverage for C10 and C12 aqueous solutions are 5.72×10^{14} and 5.41×10^{14}, respectively. The values are similar to each other. We suggest that these two concentrations lead to a saturation point or a full monolayer coverage. However, more points are needed to make that a valid assumption. Thus, a future study is considered and not yet regarded in the scope of this work. Comparing the C10 and C12 surface excess/coverage values versus the intensity ratios, with the assumption a full coverage is obtained and with an increased number of gauche defects, this scenario suggests that the interfacial molecules reached equilibrium and will only fluctuate due to other factors such as chemical nature of the surfactant, sample preparation, and equilibration time.

Another factor for the possible observation of gauche defects is that the 30-minute equilibration time or waiting time was not enough to form well-organized surfactants at the air-liquid interface at 8 mM solution. Therefore, we prepared 8 mM of C8 Quat surfactant in water and acquired ssp SFG spectra at 2900 cm⁻¹ IR center from 0, 0.5, 2, 4, to 6 hours waiting time to monitor the change in the CH spectra profile and correlated it to a peak intensity difference between CH_2 SS and CH_3 SS (**Figure B.3**). The zero (0) time means the spectrum was collected right away after placing the C8 sample at the stage. As shown in the SI, the 0-time spectral profile still has distinct peaks of the head group positioned at ~ 2972 cm^{-1} and ~3065 cm^{-1}. The ratio between the CH_2 SS and CH_3 SS slightly changes from 0-hr to 6-hr equilibration time as well, but there is no considerable effect towards the formation of gauche defects as a function of time. The results do not directly explain the nonlinear trend between the intensity ratio and the number of carbon atoms in the alkyl chain. A similar trend of nonlinear dependence between the number of gauche defects and length of the alkyl chain was also acquired as 1% and 10% of sodium chloride was introduced in the 8 mM Quat solutions (**Figure 4.3b**). Since no trend was also obtained by adding salt, this means that increasing the ionic strength or the screening effect does not straightforwardly affect the number of defects.

The overall observation obtained from the number of gauche defects and increasing alkyl chain can be supported by the explanation reported by Baldelli and Richmond groups wherein shorter chains usually possess less geometric gauche defects and less steric hindrance than longer alkyl chains.[64, 69] These Quats are highly soluble in water and their critical micelle concentration (CMC) below C10 is challenging to determine using surface tension. So, the 8 mM concentration is not enough to achieve the highest surface concentration and confined conformational mobility with the nearest neighbors from C4 to C8 as explained by the surface excess calculation results.[64, 69] Longer chain Quat compounds greater than C12 are found moderately soluble in water and thus not within the scope of this study.

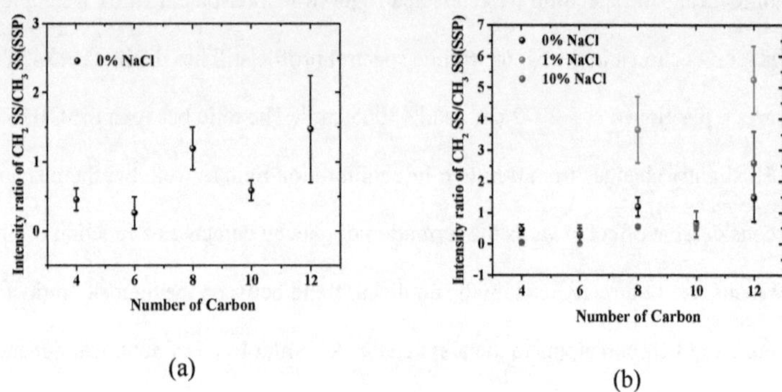

Figure 4.3. The intensity ratio of CH_2 SS at ~2851cm^{-1} and CH_3 SS at ~2876 cm^{-1} as a function of number of carbons in hydrophobic chain (a) 0% NaCl and (b) 0 %, 1 %, and 10 % NaCl.

Is the conformation of the interfacial water dependent on the length of the C_4-C_{12} Quat surfactants?

Figure 4.2 also shows that the SFG signal of OH stretches from 3000-3600 cm^{-1}. The spectral profile for ssp polarization combination with a broad peak positioned at ~3182 cm^{-1} indicates ordered water molecules at the air-liquid interface forming tetrahedral molecular arrangement in the presence of cationic Quat surfactants. The overall SF intensity of OH stretch increases from a shorter alkyl chain to a longer chain (C6 to C12) in the ssp polarization as shown in **Figure 4.4**. However, the SFG intensity of the OH stretch in the C4 Quat surfactant has a higher relative value compared to the C6 molecule. A study performed by Singh and co-workers on the dependence of structural and orientation transformations of water with varying chain lengths of alcohol at the air-liquid interface reported that the number of carbon atoms of linear chain alcohol from 1-3 (n<4) does not affect the H-bonding capability and orientation of water molecules and the interface is unstable at n=4.[101] On the other hand, when n>4, the interfacial water becomes more strongly H-bonded which was indicated by the maximum peak positioned at ~3200 cm^{-1}. As shown in **Figure 4.4**, C4 Quat behaves differently from C6 –C12 surfactants. After waiting for 30 minutes or more and performing the experiment multiple times, C4 alkyl chain length adsorbs similarly at the air-liquid interface when compared to longer chain length (C8 –C12). This indicates a structural and/or orientational transition of the interfacial water at the surfactant-water interface which was the same observation obtained by Singh and colleagues.[101] Also, according to Nguyen and colleagues, by connecting our surface tension and SFG results, the formation of this immersed surfactant bilayer results in an unaffected topmost surfactant monolayer and a reduction of surface tension value when the confined interfacial molecules re-arranged.[97] The spectral profile of the alkyl chain of the C4 molecule remains similar to other chain lengths, while the OH region showed the formation of ordered hydrogen-bonded water molecules. In addition, the transition is associated with the hydrophobic size of the chain relative to the size of the cationic headgroup. To support our explanations, we also check the pH of all solutions (C4 – C12) without salt to make sure that there is no pH dependence. The pH values for 8 mM Quat solutions are available in **Table B.3** of the Supporting Information. Since the pH values for each Quat aqueous solution are similar, the pH will globally affect the behavior of all solutions similarly. The pH value of ~5.54 obtains an acid dissociation constant (K_a) of 1.194×10^{-09}.

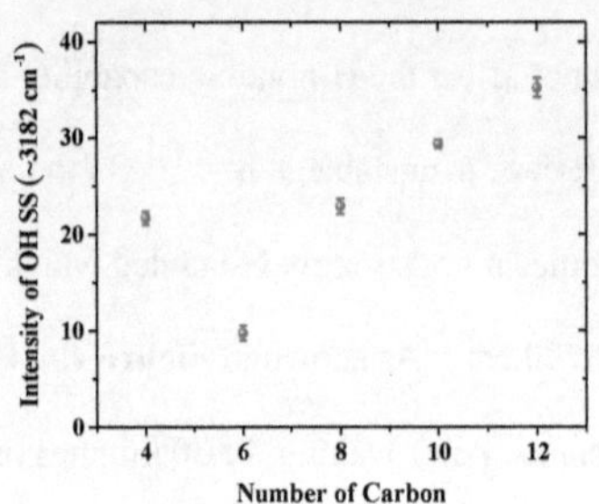

Figure 4.4. The intensity of OH SS at ~3182 cm^{-1} as a function of the number of carbons in the hydrophobic chain (0% NaCl).

Next, as shown in **Figure B.34**, 1 mM concentration of alkyl tetramethylammonium bromide (ATAB) surfactants with alkyl chain lengths of n = 4 to n=12 (4, 6, 8, 10, 12) solutions were prepared. This experiment was performed to check whether the headgroup affects the arrangement of water molecules at the air-liquid interface, especially for the C4 surfactant. However, after analysis, the ssp SFG spectrum of the C4 TAB solution did not show any evident vibrational signatures from -CH$_3$, -CH$_2$, or the -N-CH$_3$ functional groups (**Figure 4.5**). Therefore, as a result, the structural changes of interfacial water molecules observed for the C4 surfactant can be dependent on the chemical nature of the cationic headgroup. However, another possible reason for this observation could also be because the concentration of the C4 TAB solution is not enough to cause the surface activity of these C4 TAB molecules at the air-liquid interface. This means that the surface coverage of an 8 mM Quat aqueous solution is not equivalent to the surface coverage of a 1 mM C4 TAB aqueous solution. To further investigate the concern over the coverage, surface tension measurements were performed for both C4 Quat and C4 TAB solutions with a concentration range from 0 to 120 mM. This enables the comparison of the surface tension values to find the matching concentration of the 8 mM C4 Quat from the plot of C4 TAB (**Figure B.35**). The equivalent concentration was found to be ~102 mM of C4 TAB solution which will have similar surface coverage with 8 mM of C4 Quat solution. Then the ssp SFG spectra of 8 mM C4 Quat and 102 mM C4 TAB solution were acquired. **Figure 4.5** shows the spectral profiles of the two solutions. The OH stretch vibrational mode positioned at ~ 3182 cm^{-1} is evident in the ssp spectrum of the 8 mM C4 Quat solution while not clearly present in the ssp SFG spectrum of ~ 102 mM C4 TAB solution. The possible reasons for these observations are 1) the headgroup has a role in the organization of the water molecules at the air-liquid interface and/or 2) the approach of finding the matching concentration for 8 mM C4 Quat solution using the surface tension measurement does not define a 1:1 correlation with the SFG measurements. Looking closely at **Figure 4.5b**, the SSP spectrum of the 8 mM C4 Quat has the vibrational signature of the Quat headgroup positioned at 3068 cm^{-1}. This peak is a CH aromatic stretch of the benzyl group. The comparison of ssp spectra (**Figure B.36**) of the C4 Quat with 0%, 1%, and 10% shows that the presence of 1% and 10% salt has effectively reduced the spectral interference from water molecules and only emphasized the contribution of the headgroup at ~3068 cm^{-1}. Therefore, the surface activity of the C4 Quat is also assisted by the headgroup's existing interaction and overall conformation in an aqueous solution. Regarding the 1:1 correlation of the Quat and TAB surfactants concentration for surface coverage equivalence, finding its absolute correlation will require further investigation which will be performed in our future studies.

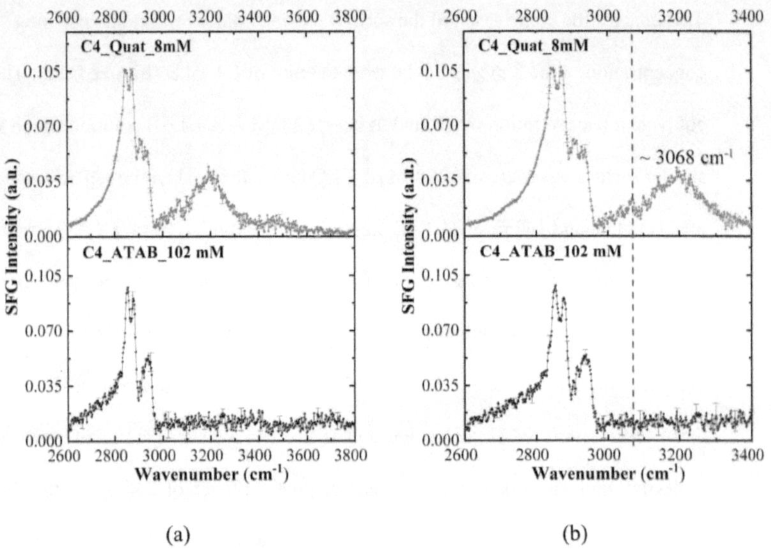

Figure 4.5. (a) ssp SFG spectra of both 8 mM C4 Quat and C4 TAB aqueous solutions from 2600 -3800 cm⁻¹ and (b) ssp SFG spectra from 2600 – 3400 cm⁻¹ to emphasize the peak positioned at ~3068 cm⁻¹.

This study also supports earlier findings that the SFG signal of the OH stretches in the 3000-3600 cm⁻¹ region are enhanced in the presence of surfactants.[62, 85, 88, 102] The adsorption of surfactants at the air-water interface creates a large electrostatic field and interfacial water molecules orient themselves relative to the surface charge of surfactants which enhances the SFG signal at the OH region. According to the flip-flop model of interfacial water molecules suggested by Tahara and co-workers, the oxygen of water molecules is pointed up to the positively charged surface. On the other hand, the hydrogen of water molecules pointed up to the negatively charged surface.[85] Using the flip-flop model of the charged surfactant-aqueous interface, the cationic Quat compound orients OH bands of interfacial water molecules with their hydrogen pointing down from the positive charge of the Quat headgroup. This study also supports the recent findings by Singh and co-workers[101] from which the long hydrophobic alkyl chains of alcohol affect interfacial water differently than the shorter chain length alcohol. However, according to their findings, the shorter chain length of alcohol does not affect the hydrogen bonding and the orientation of the interfacial water substantially and the interface is quite unstable.[101] Another group found that the adsorption of short-chain alcohol at the air-water interface is similar to the classical surfactants.[103] Critical aggregation concentration (CAC) of alcohol and the critical micelle concentration (CMC) of surfactants were considered for comparisons. The alcohols and surfactants possess linear dependence of Gibbs standard free energy of their adsorption at air-water interface as a function of hydrophobic chain length.[103] Therefore, as shown in **Figure 4.4**, the OH stretch intensity drops (~3182 cm^{-1}) from C4 to C6 can also be explained by which the interface becomes unstable because of the reduced interaction between these shorter chain length surfactants and water molecules. Thus, resulting in more randomly oriented interfacial molecules affecting the overall SFG signal of the C6 surfactant in water. In addition, the major increase of OH stretch intensity for the longer chains is direct evidence of the effect of alkyl chain length on the orientation of the interfacial water molecules and increased hydrophobic interaction between the chains. In a bulk study, Raman multivariate curve resolution (Raman-MCR) spectroscopy showed that longer chain length surfactants form micelle in the bulk and the interfacial water molecules can penetrate much deeper into the hydrophobic core of the micelle. This observation has resulted in more ordered interfacial water molecules.[104]

How do the conformations of both the chain and the water molecules change with increasing ionic strength?

Crude oil can contain very few species to numerous species inside the oil and gas pipelines. Crude oil contains dissolved salts and the concentration was reported up to (>10 wt %). NaCl is the most abundant among these salts and acts as a corrosive medium.[9, 105] The effects of salts in aqueous solutions play an important role in the aggregation behaviors of ionic surfactants.[106] The addition of salt reduces the electrostatic repulsion between the cationic headgroups, which results in their much closer assembly to each other at the air-liquid interface. This also leads to the reduction of the CMC value as well.[93] As shown in **Figures 6** and **7**, the addition of 1% and 10% salt in 8 mM aqueous solutions of C4, C6, C8, C10, and C12 Quats affected the conformation of the chain length. Looking closely at the results of both ssp and ppp polarization combinations, the changes in the intensity and spectral profiles of Quat molecules from C4 – C12 are contributions from the interactions existing between non-polar hydrocarbon (HC) chains and polar head group - water molecules due to the increasing presence of salt. The SFG spectra from a 1% NaCl solution show the signal from the chain (CH_2 and CH_3 groups), the head group (-N-CH_3 and CH aromatic groups) and –OH SS of water molecules (**Figure 4.6**).

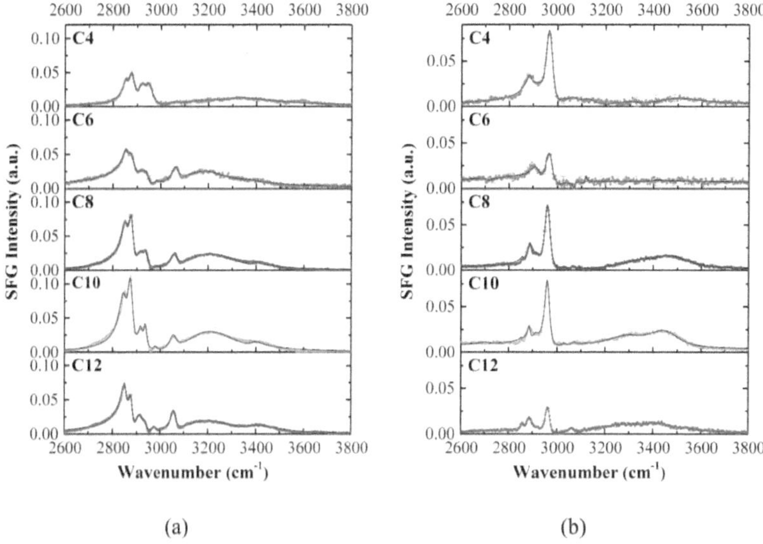

Figure 4.6. Fitted SFG spectra of 8mM C4, C6, C8, C10, and C12 in water (1% NaCl) at (a) ssp and (b) ppp polarization combinations.

Yet, with the increasing number of carbons, a decrease in the OH SS (at ~3182 cm^{-1}) is observed relative to the intensity of the –N-CH$_3$ group while the OH SS at ~3384 cm^{-1} seems to be unperturbed. For 1% (0.17 M) salt (**Table B4**), it is difficult to obtain a clear trend for the CH spectral profile, however, the vibrational signature from the headgroup becomes more distinct as the signal from the OH stretch decreases. Also, the weakly-bonded H$_2$O molecules become more prominent relative to the strongly-bonded H$_2$O molecules. Interestingly, the changes in the spectral profiles of C4, C6, C8, C10, and C12 in the presence of 10% salt (1.7 M) are more noticeable (**Figure 7**). SFG signal from interactions between the alkyl chains and arrangement of the head group is still

noticeable from C4- C12 Quat surfactants. However, the SFG signal from the strongly-

bonded water molecules positioned at ~3182 cm^{-1} had decreased when compared to the

results obtained from 8 mM Quat aqueous solution with 1% NaCl. Thus, a lesser

contribution was evident from the constructive interference created between the OH and

CH vibrational modes. The CH spectral profiles show no trend, yet carefully analyzing

the distinct C4 ssp and ppp spectra, the intensity ratio between CH_2 SS and CH_3 SS

decreases. This means less number of gauche defects with 1% NaCl. In addition, we also

calculated the tilt angle [50] for the terminal methyl group of C4 and found the value to be

$39° \pm 27°$ for 0% salt and $45° \pm 84°$ for 1% salt from the surface normal. This

observation regarding gauche defects is due to the addition of salt which reduces the

electrostatic repulsion between the head groups, allowing the close packing of the

surfactants at the air-liquid interface. Therefore, more surfactants are adsorbed at the

interface which facilitates the decrease in its surface tension as well as reduction of the

electrostatic field. The hydrophobic interactions between alkyl chains arranged the polar

head group in an orderly manner to each other. At the same time, the OH group

contribution has been reduced affecting the arrangement of the water molecules in the

double-layer region via formation of hydrogen bonds along with some contributions from

the intramolecular interaction as a result of the vibronic coupling of the OH SS with the

bending mode.[63, 72] To summarize, **Figure 4.8** shows the graph with OH SS at ~3182 cm^{-1}

and ~3384 cm^{-1} from 0%, 1%, and 10% as a function of increasing chain length.

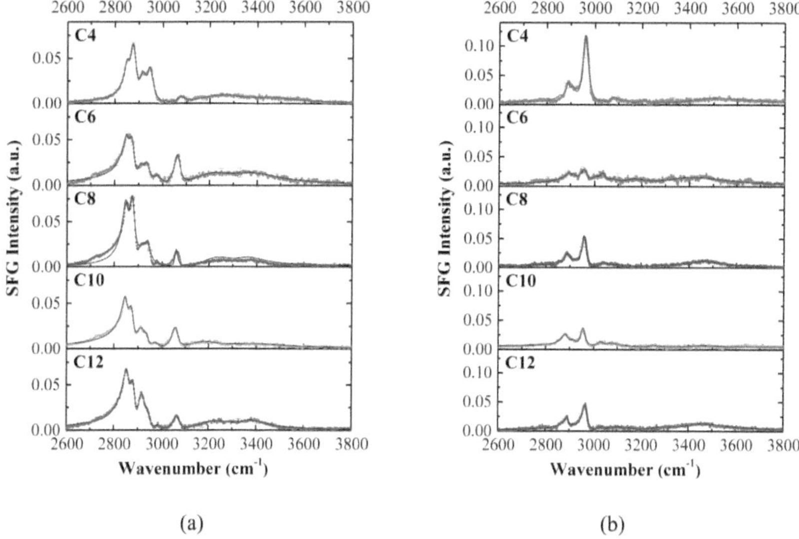

Figure 4.7. Fitted SFG spectra of 8mM C4, C6, C8, C10, and C12 in water (10% NaCl) at (a) ssp and (b) ppp polarization combinations.

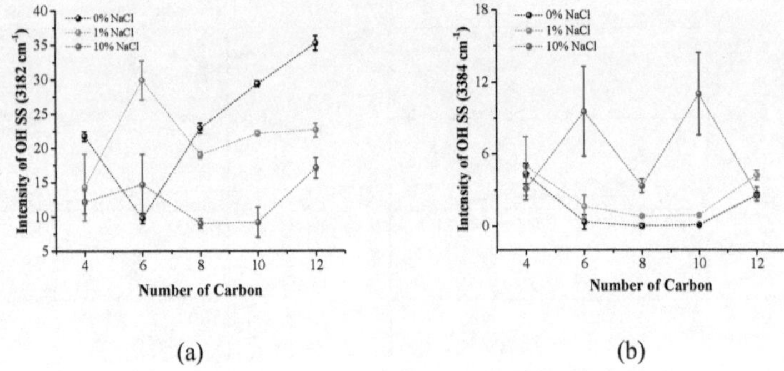

Figure 4.8. The plot of intensity of OH SS at (a) ~3182 cm^{-1} and (b) ~3384 cm^{-1} as a function of number of carbon atoms in the alkyl chain at different NaCl concentrations.

However, the overall effect introduced by salt in the solutions of increasing chain length does not provide a clear trend. Therefore, in **Figure B.7**, the OH stretch was plotted as a function of 0, 1, and 10% salt concentration for C4 to C12 surfactants. There is no clear trend for C6 Quat because, at 0%, the OH intensity is low and increases at 1% and then decreases again for 10%. However, a decreasing trend of the OH intensity at ~3182 cm^{-1} is observed from C4, C8, C10, to C12. Consequently, the OH stretch signal at ~3182 cm^{-1} decreases as we increased the salt concentration from 0, 1% to 10%. The implication of can be very well-correlated to the affected contributions of the second-order nonlinear susceptibility, $\chi^{(2)}$, and the third-order optical properties of the bulk water, $\chi^{(3)}$, in view of the overall SFG signal from a charged interface and decreasing OH stretch interference.[72] The C6 Quat molecule in water was characterized multiple times to ensure the repeatability of the spectral results. In such a case, we must find a reason behind the C6 response to increasing salt concentration upon monitoring the OH stretch at ~3182 cm^{-1}. The hydrophobic size of the C4 Quat compound may have affected the OH stretch intensity, which resulted in a structural transformation of the interfacial water and that is why its OH spectra profile is similar to C8, C10, and C12. On the other hand, the H$_2$O molecules are not strongly hydrogen-bonded in the aqueous solution of C6 Quat molecules with 0% salt. At 1% salt, an increase in the SFG signal was obtained at ~3182 cm^{-1}, which indicated that the H$_2$O molecules at the interface and the double layer are arranged more orderly. Then, a sudden decrease for the OH SS vibrational mode SFG signal was observed at 10% salt which resulted in a decrease in the order between the arranged H$_2$O molecules.

In the case of the shortest chain length (C4), our results indicated a more ordered monolayer was achieved by increasing the NaCl concentration to 1.70 M (**Figure B.36**). This increased contribution from the terminal methyl symmetric suggested a considerable effect on the ordering of short-chain surfactants due to ionic strength when compared to the presence of no salt and ~1.70 M salt. In addition, increasing the ionic strength results in a decrease in the OH band of the interfacial water at a positively charged interface.[74] As shown in the C4 spectra from 0%, 1%, and 10% salt content (**Figure B.6**), the OH band signal is relatively reduced where water ordering is random; yet has a more ordered hydrocarbon chain.

How does increasing ionic strength specifically affects the OH band?

In this discussion, 8 mM C8 Quat molecule was prepared with 0%, 0.5%, 1.0%, 5%, and 10% salt, which has equivalent /corresponding ionic strength values listed in **Table B.4** available in the Supporting Information. **Figure 4.9a** presents ssp spectra for the C8 Quat molecules and **Figure 4.9b** plots the intensity of the OH band positioned at

~3182 cm^{-1}. The ppp spectra are shown in the Supporting Information (**Figure B.38**). The

OH band decreases with increasing ionic strength. This observation, as noted earlier, is

because of the screening effect of the increasing ionic strength. Also, a decrease in the

overall SFG signal in the CH region is also a consequence of adding the salt. The

addition of 0,5% and 5% of NaCl relatively shows the gradual decrease in the OH SS at

~3182 cm^{-1}.

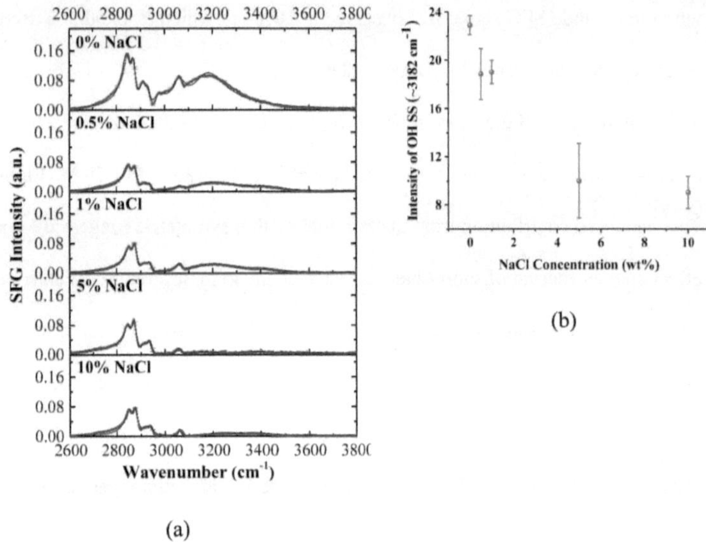

(a)

Figure 4.9. (a) Fitted SFG spectra of 8mM C8 in water with 0%, 0.5%, 1%, 5%, 10%

NaCl at ssp Polarization combinations and (b) the intensity of OH SS at ~3182 cm^{-1} as a

function of NaCl concentration.

Conclusion

The chain length dependence of C4, C6, C8, C10, and C12 Quat surfactants have been successfully studied with different ionic strengths at the air-water interfaces by SFG. From the SFG spectra, we can conclude that this series of surfactants retain surface activity with considerable number of gauche defects even at short alkyl chain length. The previous studies focusing on chain length and ionic strength reported their observations about gauche defects of alkyl chain[39, 43, 68, 69, 101] and conformational changes of interfacial water independently.[50, 62, 63, 73, 74, 101] The number of gauche defects decreases with increasing chain length was observed at air-solid and air-liquid interfaces[67, 68] and decreasing trend of gauche defects were reported with decreasing chain length at liquid-liquid interface.[69] On the other hand, the ionic strength dependence shows that the interfacial water molecules become less ordered with increasing salt concentration.[50, 62, 72, 73] In this work, we reported the combined effect of chain length and ionic strength variations to the conformational changes of surfactant-water interfaces. This research provides several new findings: (1) No clear trend was obtained for the number of gauche defects as a function of increasing chain length from C4 – C12. Less number of gauche defects were observed for shorter chain lengths while an increasing number of gauche defects were observed for longer carbon chains, (2) The number of gauche defects decreases for the shortest chain surfactant (C4) as a function of ionic strength, (3) Increasing the alkyl chain length (C6, C8, C10, and C12) results to more ordered interfacial water molecules. The OH SS vibrational mode of C8 Quat at 3182 cm^{-1} in water has decreased with increasing ionic strength, (4) At a similar surface coverage, C4 Quat showed SFG signal from OH SS vibrational modes as a signature for more ordered water molecules at the air-liquid interfaces compared to a C4 TAB surfactant. While, no evident SFG signal from the OH vibrational mode was observed for C4 TAB surfactant. This result showcases the role of the chemical nature of headgroup towards the formation of tetrahedrally arranged water molecules at the air-liquid interface in the presence of a shorter alkyl chain, and (5) The hydrogen bonding formation between water molecules was affected and enhanced in the presence of a C4 Quat surfactant. On the other hand, a recent study of alcohols with varying alkyl length reported that the H-bonding formation is unaffected with a C4 alcohol.[101] These findings support previous studies on chain length[39, 43, 68, 69, 101] and ionic strength.[50, 62, 63, 73, 74] This work provides an evidence to consider effective number of carbon on alkyl chain at different ionic strength, self-assembly of surfactants with different headgroup, and conformational changes of interfacial water molecules undergone by increasing chain length and ionic strength which leads to our future work on self-assembly of surfactants at liquid-metal interface. In addition, we presented the adsorption behavior of a series of highly water soluble short chain surfactants of limited CMCs which will benefit in designing new surfactants for a specific environment in corrosion inhibition,[62-67], fabric softeners, interfacial processes, self-assembly, micellization, and other related fields.[22, 61, 107, 108]

CHAPTER 5: DIRECT OBSERVATION OF ADSORPTION MORPHOLOGIES OF CATIONIC SURFACTANTS AT THE GOLD METAL-LIQUID INTERFACE

Introduction

Internal corrosion of oil-and-gas pipelines is a major health, safety, and environmental problem that is estimated to cost roughly $ 7 billion/yr in the United States alone.[2] Corrosion inhibitors are surfactant molecules that are injected into the oil-and-gas pipelines to retard corrosion.[109] Surfactants are amphiphilic compounds. Due to their chemical structure, surfactants tend to adsorb onto solid metal surfaces from an aqueous solution. These surfactants are understood to create a hydrophobic barrier for corrosive substances like carbon dioxide (CO_2), hydrogen sulfide (H_2S), and water (H_2O), which slows down the rate of corrosion.[110, 111] For effective corrosion inhibition, it is desired that the alkyl tails of adsorbed surfactants form a well-packed, ordered layer on the metal surface.[112] This raises an important question as to how do the alkyl tails affect the adsorption morphologies of the surfactants. Previous studies of the adsorption of anionic and cationic surfactants on metal oxide surfaces suggest that the initial adsorption is driven by the affinity between the charged head groups of the surfactants and the surfaces upon application of a stimulus (concentration and pH changes). [113, 114] Atomic force microscopy and surface-enhanced Raman spectroscopy have also revealed the presence of ordered layers of adsorbed surfactants on metal surfaces, which suggests that hydrophobic interactions between the alkyl tails play a role in the adsorption process.[115, 116] However, there has not been any direct measurement of the molecular configurations in the adsorbed layers. Other studies have examined the adsorption behavior of monolayers on a metal surface at the air-solid and liquid-solid interfaces facilitated by an applied potential.[117-121] But the spontaneous adsorption of surfactants from aqueous solution to the metal surface without external stimuli has not been investigated as yet.

In this report, for the first time, the *in-situ* adsorption of cationic surfactants (Quats) and their assembly, in ordered morphologies without any assisting mechanism studied via SFG spectroscopy, has been reported. SFG has been employed previously to observe pre-adsorbed monolayers before exposure to a liquid environment under electrochemical conditions[122, 123] and to also directly observe corrosion and electrochemical reduction products at the metal-liquid interface.[124, 125] Observing *in-situ* organization of surfactants on metal surfaces is challenging and predicting the adsorption behavior is difficult because of the many interactions that are involved in the adsorption process. We have studied the adsorption and self-assembly of C4 and C12 Quat surfactants using SFG to also understand the effect of the length of alkyl tails on the adsorption behavior. In this study, we present direct measurements of molecular-level details of the adsorbed layers of C4 and C12. Along with SFG, we have also performed

fully atomistic simulations of C4 and C12 at the gold-water interface.

SFG Background

SFG spectroscopy is an interface-selective technique that has been popularly used

to obtain vibrational spectra of monolayers formed on metal and dielectric surfaces at a

molecular level.[120, 126-128] The spectral analysis provides information about conformation,

chemical identification, orientation, and dynamics of adsorbed interfacial molecules.[28, 49]

The details of SFG theory and setup have been described elsewhere.[24, 28, 29, 49, 50, 128].

The SFG signal (ω_{SF}) generates with the temporal and spatial overlaps of broadband IR

beam (ω_{IR}) and a narrowband visible beam (ω_{VIS}) at the interface.[49, 50]

$$\omega_{SF} = \omega_{VIS} + \omega_{IR} \tag{5.1}$$

The efficiency of the SFG process is directly related to the second-order nonlinear

effective susceptibility $\chi_{eff}^{(2)}$, the combination of both resonant susceptibility, $\chi_R^{(2)}$, and the

nonresonant susceptibility, $\chi_{NR}^{(2)}$.[49]

$$\chi_{eff}^{(2)} = \chi_R^{(2)} + \chi_{NR}^{(2)} \tag{5.2}$$

The nonresonant term, $\chi_{NR}^{(2)}$, depends on the substrate surface of solid and is almost

negligible for liquid and dielectric surfaces.[49] When the frequency of the incident IR

beam is in resonance with the vibrational modes of adsorbed molecules at the interface,

the resonant term, $\chi_R^{(2)}$, becomes large and contributes in intensity of the SFG signal.[29, 129]

All the SFG spectra were collected in either ssp or ppp polarization combinations.

Herein, (s) and (p) refer to incident beam perpendicular and parallel to the plane of

incidence, respectively.[49] The ssp indicates the SFG beam is (s) polarized, visible beam is

(s) polarized, and the IR beam is (p) polarized as the order of decreasing energy and ppp indicates all three beams are (p) polarized.[50]

Experimental Details

Synthesis and Purification of Alkylbenzyldimethylammonium Bromide (Quats)

Alkylbenzyldimethylammonium bromides containing two different alkyl tail lengths butyl (C4), and dodecyl (C12) were synthesized using established methods.[47, 50] Equimolar N, N-dimethylbenzylamine, respective bromoalkane, and acetonitrile solvent were added into the three-necked round-bottom flask. After that, the solution was heated to a reflux temperature of 82°C for 24 hours. The synthesized products were purified by recrystallization process with deionized water. The structure and the purity of Quats were confirmed by ^{1}H-NMR spectra with their corresponding purity values available in earlier publications.[49, 50]

Sample Preparation

A 1 cm x 1 cm gold thin film on a silicon substrate (99.999% Au, 1000 Å thick from Sigma Aldrich, St. Louis MO) was cleaned with ethanol and dried with nitrogen gas. The cleaning procedure was followed by plasma cleaning to remove any impurities and surface contaminants. The freshly prepared Quat solution (~5 ml) in D_2O was injected into the Teflon sample cell (**Figure 5.1**) containing the gold thin film substrate. Both visible and IR beams were temporally and spatially overlapped at the liquid-gold metal interface to generate the SFG signal.

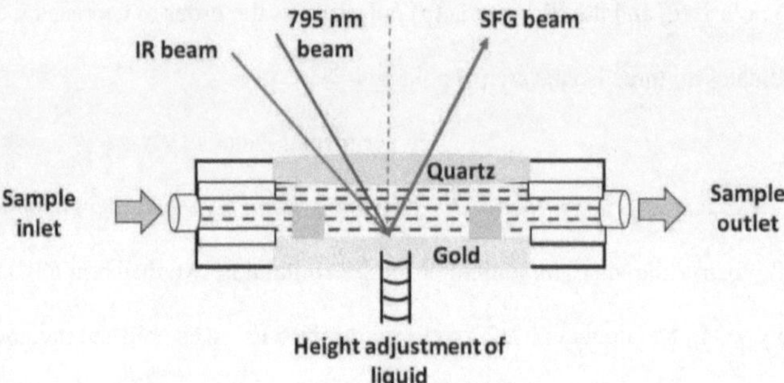

Figure 5.1: Cross-sectional cartoon of the liquid-solid sample cell. The cell contains an inlet, an outlet, a fused silica window that helps seal the cell, a Teflon spacer (~254 μm), and a sample holder with an adjustable height to maintain an exact volume of the deuterated aqueous solution between the window and the metal surface.

Data Acquisition

SFG spectra were acquired at the CH region centered at 2900 cm^{-1} with ssp and ppp polarization combinations. Each spectrum was collected for 9 minutes and corrected for background, smoothed, and fitted with the Lorentzian line shape equation shown below.[49, 50] The fitting results are available in **Table 5.1** and **Table 5.2**. The average tilt angle values of the methyl (CH$_3$) terminal group were estimated from the fitted data.[50]

Fitting Equation

The analysis of each polarization combination and the line broadening was considered by the following Lorentzian line shape equation.[49, 50]

$$I_{SFG} \propto \left| \chi^{(2)} \right|^2 \propto \left| \sum_q \frac{N\langle\beta^{(2)}\rangle}{(\omega_{IR} - \omega_q + i\Gamma_q)} + \left|\chi_{NR}^{(2)}\right| e^{i\rho} \right|^2 \qquad (5.3)$$

Where ρ is the phase of the nonresonant response. The nonresonant contribution from the gold substrate and the Gaussian equation are both included in the fitting equation to account for the mid-infrared beam's broadband pulse width. The simplified version of the fitting equation that accounts for the CH vibrational modes is presented below.

$$I_{SFG}\,(\omega + \omega_{vis}) \propto exp\left[-\frac{(\omega-\omega_{IR}^L)^2}{2(\delta\omega_L)^2}\right] \times \left|\Sigma_q\frac{A_q}{\omega_{IR}-\omega_q+i\Gamma_q} + A_{NR}e^{i\rho}\right|^2 \qquad (5.4)$$

The Gaussian curve equation is defined with the spectral width $\delta\omega_L$ centered at ω_{IR}^L. The spectral width $\delta\omega_L$ centered at ω_{IR}^L is included in the Gaussian function. The amplitude factors, A_q and A_{NR}, are proportional to the molecular hyperpolarizabilities as shown above in equation 4.[49, 50]

The adsorption behavior of alkyl dimethyl benzylammonium bromide (Quat) of two different tail lengths (n = 4 and n = 12, henceforth referred to as C4 and C12, respectively) at the gold metal-liquid interface was investigated by SFG spectroscopy. The application of the SFG technique to the water-solid interface is experimentally challenging due to the strong absorption of the infrared beam by the water molecules along with the non-resonant background coming from the metal surface. Here, we probed this interface using a geometry shown in **Figure 5.2** where the incident beams (visible and infrared) passed through the fused silica window and the liquid to the metal surface. Then, the beams including the SFG beam were reflected from the metal through the liquid and the window. The solution was sandwiched using a ~254 μm thickness Teflon spacer in between the fused silica window and the metal surface. The thickness of the spacer is sufficient to reduce interference between the window and the gold metal surface and other experimental artifacts. The systematic suppression of nonresonant contribution from the metal surface was achieved without distorting the original SFG spectra. This

approach contributes towards expanding applications of SFG to study other phenomena

at the metal-liquid interfaces. The details on data acquisition, and fitting analysis,

including the results for the determination of the average tilt angles, are shown below.

The structure of the surfactant molecules and a schematic showing adsorption of C4

molecules from the solution scheme is shown in **Figure 5.2**.

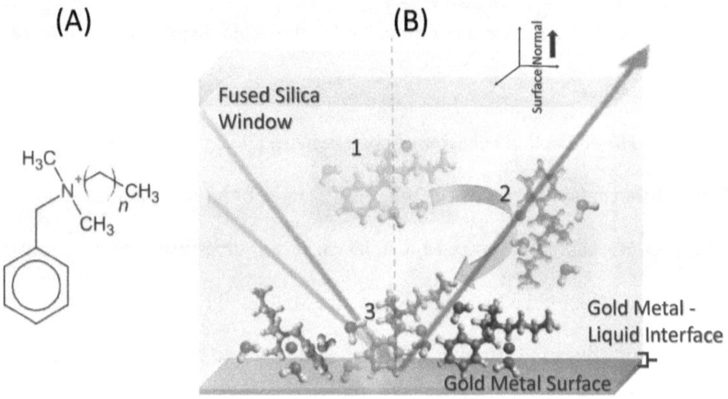

Figure 5.2. (A) Structure of Quaternary ammonium-based surfactant molecules

employed in this study. The subscript n indicates the number of CH_2 groups in the alkyl

tail (where n = 3 and n = 11 carbon atoms as C4 and C12, respectively). (B) Schematic

diagram of a C4 molecule's mechanism of adsorption from the (1 and 2) solution to the

gold metal surface (3).

C4 and C12 Quat solutions of concentration 0.4 mM in D_2O, which is below the critical micelle concentrations (CMCs) of both C4 and C12 Quat [49] were allowed to equilibrate for an hour inside the SFG cell. D_2O was used to avoid any interference from the OH vibrational modes in the analysis of the CH vibrational modes. The spectra were collected by suppressing the non-resonant contribution from the gold substrate.[130] The ppp polarization combination generates a stronger SFG signal compared to the ssp polarization combinations at the metallic surface.[25] Because of this strong NR signal from ppp polarization combination, it requires a longer time delay between the visible and IR beams to effectively suppress the background contribution. The time delay for ssp and ppp polarization combinations were ~2.40 ps and ~3.00 ps, respectively. The delays were determined by collecting the spectra at different times until the nonresonant background is sufficiently suppressed. An example of how this suppression was developed at different timing delays is shown in **Figure 5.3** for C12 Quat.

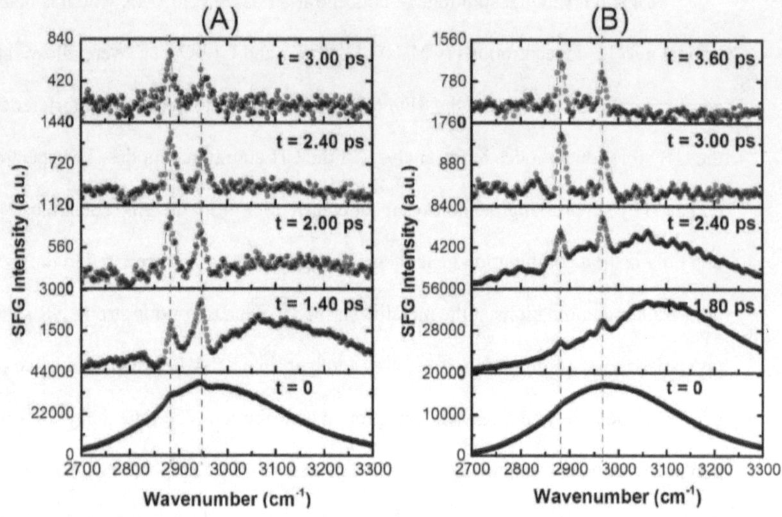

Figure 5.3. (A) ssp and (B) ppp SFG spectra of a C12 Quat acquired at different delays.

Results and Discussion

In **Figure 5.4,** the C4 Quat ssp spectrum (A) contained a prominent methylene asymmetric stretch peak (CH_2 AS) at ~2907 cm^{-1}.[131, 132] The other CH stretches are less prominent (peak assignments and positions are reported in the SI).[49, 50, 77, 78]

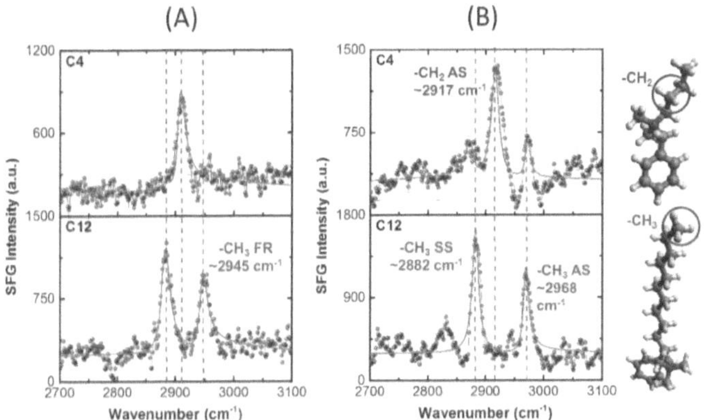

Figure 5.4. Fitted SFG spectra of 0.4 mM C4 and C12 Quat in D_2O at (A) ssp and (B) ppp polarization combinations.

The methyl asymmetric stretch was not evident in the ssp spectrum compared to other vibrational modes especially from the methylene groups that dominated the surface. In the C12 Quat ssp spectrum, sharp peaks of the terminal methyl symmetric (CH_3 SS) ($\sim$2884 cm^{-1}) and methyl symmetric stretch split by the Fermi resonance interaction with the methyl bending mode (CH_3 FR) ($\sim$2945 cm^{-1}) are observed.[78, 79] In the C4 Quat ppp spectrum (B), CH_3 SS, CH_2 AS, and methylene asymmetric stretch (CH_3 AS) vibrational modes are observed at $\sim$2880 cm^{-1}, $\sim$2917 cm^{-1}, and $\sim$2972 cm^{-1}, respectively.[77, 78, 133] In addition, the CH_3 AS present is more evident in the ppp spectrum while not evident in the ssp spectrum, the possible reason for this observation can be because most of the AS vibrational modes are not aligned along with the s-polarized lights of the visible and SF beams in the ssp polarization. The CH_3 AS is otherwise evident in the ppp spectrum. As the polarization of both incident beams is parallel to the incident plane (p-polarized), AS stretches are probed by the p-polarized visible and IR beams while only p-polarized SFG signal is recorded. This could be because CH_3 AS vibrational modes lie parallel to the incident plane.[134, 135] On the other hand, peaks of the CH_3 SS, and CH_3 AS peaks are observed and prominent in the C12 Quat ppp spectrum. [77, 79, 133] **Tables C3** and **C4** summarize the peak positions from fittings and assignments, which are available in the Appendix C.

Table 5.1. Peak assignments from the fitted SFG spectra of C4 Quat at ssp and ppp polarization combinations.

Peak assignments	ssp wavenumber cm^{-1}	ppp wavenumber cm^{-1}
Methylene symmetric stretch (CH_2 SS) [49, 50, 77, 78]	~2860	-
Methyl symmetric stretch (CH_3 SS)[49, 50, 77, 78]	~2885	~2880
Methylene asymmetric stretch (CH_2 AS)[131, 132]	~2907	~2917
Methyl symmetric stretch split by the Fermi resonance interaction with the methyl bending mode (CH_3 FR)[78, 79]	~2951	-
Methyl asymmetric stretch (CH_3 AS)[133]	-	~2972

Table 5.2. Peak assignments from the fitted SFG spectra of C12 Quat at ssp and ppp polarization combinations.

Peak assignments	ssp wavenumber cm^{-1}	ppp wavenumber cm^{-1}
Methylene symmetric stretch (CH$_2$ SS)[49, 50, 77, 78]	~2859	-
Methyl symmetric stretch (CH$_3$ SS)[49, 50, 77, 78]	~2884	~2882
Methyl symmetric stretch split by the Fermi resonance interaction with the methyl bending mode (CH$_3$ FR)[78, 79]	~2945	~2944
Methyl asymmetric stretch (CH$_3$ AS)[133]	-	~2968

Table 5.3. Orientation parameters were obtained from the simulation curves for the polarization intensity and amplitude ratios of CH_3 SS/CH_3 AS vibrational modes to estimate average tilt angles and distribution angles. The errors presented are calculated at 95% CI.

C4 Quat		C12 Quat	
Average tilt angle	51°±13°	Average tilt angle	46°± 0.1°
Distribution angles		Distribution angles	
22°±2 (for 60° tilt angle)		27°±0.03 (for 50° tilt angle)	
27°±3 (for 70° tilt angle)		34°±0.04 (for 60° tilt angle)	
29°±4 (for 80° tilt angle)		37°±0.04 (for 70° tilt angle)	
		38°±0.04 (for 80° tilt angle)	

The C12 spectrum in **Figure 5.4** is dominated by CH_3 vibrational modes while the CH_2 modes are barely noticeable. This observation indicates that the C12 alkyl tail is ordered because the tails are arranged in trans-conformation. As a result, the tails have a close-packed arrangement due to the increased hydrophobic tail-tail interaction and initial adsorption driven by the affinity of the charged headgroup to the gold metal surface.[113, 114, 136] In contrast, the C4 spectrum is dominated by the CH_2 mode and the CH_3 modes are less visible. This observation indicates that the C4 alkyl tail is less ordered; resulting in a random orientation and more gauche defects due to weaker hydrophobic interactions between the shorter tails.[137] However, Coulombic interactions between the C4 surfactant and the metal surface are still significant enough that the C4 Quat surfactant is observed at the gold metal-liquid interface. Also, the average tilt angle (θ) values of the methyl (CH3) terminal group of C4 and C12 Quat (**Figure 5.5**) were estimated from the fitted data.[50] The θ is the angle of the terminal methyl group from the surface normal. The θ values were determined by taking the intensity ratios between the CH3 SS and CH3 AS from the fitted ppp spectra.[49, 50] Then, the intensity ratios must be intersected with the SFG simulated curve to estimate the θ (**Figure 3**). [29] The θ values are 51° ± 13 and 46° ± 0.1° for C4 and C12, respectively. θ values of C4 show a large standard deviation indicating that the molecules are not ordered unlike in the case of C12 molecules or also due to the goodness of the fit. The smaller deviation in the tilt angle of C12 is a consequence of stronger hydrophobic effects between the tails of the surfactants, which leads to an ordered structure. In addition, we are also reporting the distribution angles for both C4 and C12 Quat molecules. The distribution angles for C4 are 22° ± 3° (60° tilt angle), 27° ± 3° (70° tilt angle), and 29° ± 4° (80° tilt angle) whereas for C12 we have the following values 27° ± 0.03° (50° tilt angle), 34° ± 0.04° (60° tilt angle), 37° ± 0.04° (70° tilt angle), and 38° ± 0.04° (80° tilt angle). This also means that the distribution angles estimated for every tilt angle are narrow, ranging from 22° to 38°, which is less than 90° (defined as a broad distribution).[49, 50, 134, 138, 139] **Table C6** summarizes the average tilt angles and distribution angles for the methyl groups of C4 and C12 Quats.

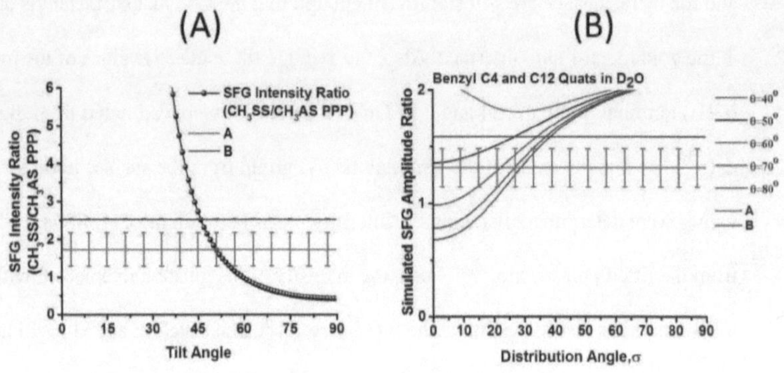

Figure 5.5. (A) The fitted SFG intensity ratios (CH₃ SS/CH₃AS ppp) are matched to the SFG simulated curves as a function of the tilt angle to estimate the average tilt angles and (B) the fitted SFG amplitude ratios (CH₃ SS/CH₃AS ppp) are also matched to the SFG simulated curves as a function of the distribution angle (σ) to estimate the distribution of tilt angles for C4 and C12 Quat molecules.

The adsorbed configurations of C4 and C12 Quat molecules were studied via MD simulations. Following previous works[140, 141], we initiated the simulations by organizing the surfactants in a SAM configuration at the gold metal-water interface (**Figure C3, Appendix C**). Canonical ensemble MD simulations were performed with a vapor space of 20 Å to ensure that the system is maintained at saturation pressure.[142] The simulations were run for 580 − 773 ns. Equilibration was ensured by comparing the ensemble-averaged distribution of surfactants, water, and counterions of successive time-periods of

40 ns. Details of the simulation set-up are available in the **Appendix C**. **Figure 5.6** shows configurations of the C4 and C12 Quat molecules after the MD simulations.

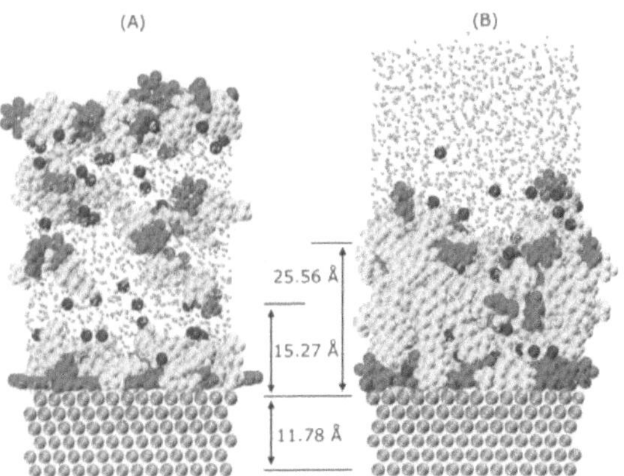

Figure 5.6. Snapshots of the configuration of (A) C4 molecules, and (B) C12 molecules at t = 677 ns and t = 500 ns respectively. Red beads represent the aromatic rings, and the yellow beads represent the alkyl tails. Water molecules are represented in cyan color and the gold lattice is represented by orange beads. The blue beads represent bromides. 15.27 Å is the length of a C12 molecule.

A distribution of tilt angles is also computed from MD simulations and is shown in **Figure C4** (Appendix C). The average tilt angles are found to be 59.3° ± 1.4° and 52.1° ± 1.0° for the C4 and the C12 molecules, respectively. The values of average tilt angles obtained in the MD simulations are in qualitative agreement with the estimated average tilt angles from SFG spectroscopy.

Figure 5.6(A) shows the distribution of orientation of alkyl tails of the surfactants with respect to the surface normal, averaged over the last 120 ns of simulation. We have calculated the orientation of the alkyl tails by first defining the end-to-end vector as a vector joining the nitrogen atom of the polar head to the carbon atom of the terminal CH_3 group. The orientation profiles show that a fraction of C4 Quat molecules adsorbed with their molecular axis perpendicular to the surface normal. The distribution profile is calculated as $\frac{\langle H(\theta)\rangle}{\langle N\rangle sin\theta}$ wherein $\langle H(\theta)\rangle$ is the ensemble-averaged histogram of the angle θ between the end-to-end vector of the adsorbed molecules and the surface normal; and $\langle N\rangle$ is the ensemble-average number of adsorbed molecules. The distribution profile shows that C12 Quat molecules are predominantly aligned close to the surface normal. On the other hand, the distribution of C4 Quat molecules has a peak at ~90° and a broad distribution around 25°. This suggests that while the adsorbed C12 Quat molecules form an ordered configuration, the adsorption of C4 molecules is disordered. These MD results support the observations for C4 and C12 molecules using SFG.

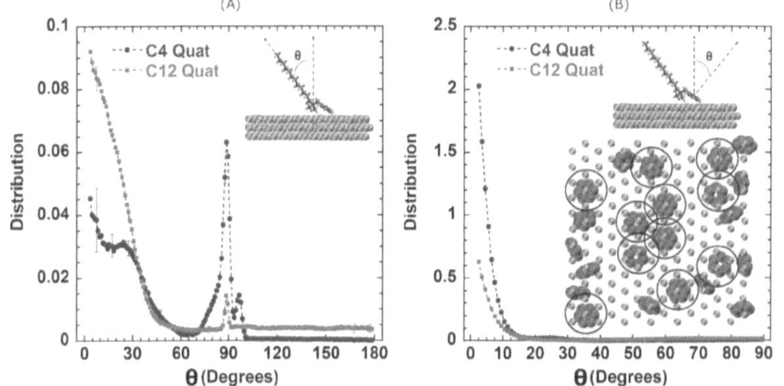

Figure 5.7. Distribution of orientation of (A) alkyl tails, and (B) normal vector of the aromatic rings with respect to the surface normal of adsorbed C4 and C12 molecules. Inset in (B) shows the aromatic rings of C12 Quats that are close to the surface. A good fraction of them lies parallel to the surface. Error bars in (B) are smaller than the size of the markers.

Figure 5.6(B) shows the distribution of the angle that the normal vector of the aromatic rings form with the surface normal. This distribution is also calculated as $\frac{\langle H(\theta) \rangle}{\langle N \rangle sin\theta}$ where $\langle H(\theta) \rangle$ is now the ensemble-averaged histogram of the angle between the normal vector of the aromatic rings of adsorbed molecules and the surface normal. A peak in the distribution at $\theta \approx 0°$ implies that in the adsorbed state of both C4 and C12 Quat molecules, the aromatic rings predominantly lie flat onto the surface. An important conclusion from these distributions is that the alkyl tails of C12 molecules mostly stand-up on the surface. This is further confirmed from **Figure C5 (Appendix C)** which shows the distribution of the angle between the alkyl tail and the normal vector to the aromatic ring of the adsorbed molecules. Analogous to **Figure C5, Appendix C** is **Figure C6, Appendix C**), which shows the same distribution, but for the molecules in the bulk aqueous phase. Most adsorbed C12 molecules are bent as compared to their conformation in the bulk phase. We have confirmed the validity of our simulation results by studying a larger simulation system with twice the surface area and the number of C12 molecules (**Figure C10, Appendix C**). Overall, our MD simulation results verify the experimental findings that the C12 molecules form a well-ordered adsorbed configuration, whereas the C4 molecules are adsorbed in random orientations. Also, the MD simulations reveal details about conformations of the adsorbed molecules in the ordered monolayer.

Conclusions

In summary, by performing interfacial analysis using sum frequency generation (SFG) spectroscopy, we have studied the unassisted adsorption behavior of surfactants onto the gold metal surface from an aqueous solution. This study demonstrates that the *in-situ* SFG spectroscopy is a powerful tool to probe the adsorption and self-assembly of cationic surfactants and perform orientational analysis at the liquid-solid interface. The longer alkyl tail surfactant (C12) formed an ordered monolayer onto the metal surface and has trans-conformation because of strong tail-tail interactions. However, the shorter alkyl tail surfactant (C4) formed a less ordered or randomly oriented monolayer on the metal surface due to the weak tail-tail interactions. We attribute these findings to tail lengths, conformation, and adsorption as key factors for SAM formation. Our simulation results corroborate with experiments, showing that the C4 molecules adsorb in random orientations, whereas the C12 molecules adsorb in a well-packed ordered monolayer configuration. These fundamental findings will also lead to our future work on the adsorption of surfactants onto the commercial-grade mild steel at different ionic strengths or salt concentrations with and without the applied potential to mimic the actual corrosive environment. This systematic study will facilitate the design of better corrosion inhibitors for specific corrosive environments. These findings are important for studying liquid-solid interfacial phenomena such as corrosion, and for designing better corrosion inhibitors for oil-and-gas transportation pipeline industries.

CHAPTER 6: SELF-ASSEMBLY OF C12 QUAT AT IRON-D$_2$O INTERFACE

Introduction

Surfactants are known to adsorb onto metal surfaces from aqueous solution or gaseous phase and self-assemble in organized structures.[57, 136] The self-assembly process is understood to be due to Coulombic interactions between the cationic head group and the metal surface, as well as hydrophobic interactions between the alkyl tails.[113, 114] The hydrophobic interactions between the alkyl tails can play a critical role in the formation of self-assembled monolayers (SAMs) by promoting their packing on the metal surface.[136] The formation of adsorbed SAMs is vital for various applications, such as corrosion inhibition, friction reduction, catalysis, wetting, micro and nanofabrication, biological membranes, cellular structure, and molecular recognition.[33, 136, 143-146] SAMs' structure, packing density, and molecular orientation have been extensively investigated by atomic force microscopy (AFM), scanning tunneling microscopy (STM), quartz crystal microbalance (QCM), X-ray diffraction, infrared (IR), Fourier transform infrared spectroscopy (FTIR), and infrared reflection/absorption spectroscopy (IRRAS), including sum frequency generation spectroscopy (SFG).[17, 33, 49, 50, 145, 147, 148] SFG has been used to observe a pre-adsorbed monolayer onto the metal surface before exposure to a liquid environment under electrochemical conditions[122, 123] and to also perform direct observation of corrosion and electrochemical reduction products at the gold metal-liquid interface.[149, 150]

In this chapter, we report, the *in-situ* adsorption of cationic surfactants (Quats) onto the Fe metal surface from an aqueous solution by SFG spectroscopy. SFG spectroscopy at the liquid-metal interface is experimentally challenging due to the interaction of different interfaces and inherent intense nonresonant SFG signal associated with the metal substrate.[57]

Experimental Methods

Iron (Fe) Substrate Preparation

(1 inch x 1 inch x 0.0300 inch) was purchased from Goodfellow. At first, the oxidized Fe surface was polished with 1200 grit sandpaper. Then, the Fe piece was further polished with 1200 grit emery paper. Next, the Fe piece was polished using diamond paste (Electron Microscopy Sciences) and aluminum slurry (Electron Microscopy Sciences) in nylon cloth (Electron Microscopy Sciences). The Fe substrate

was sequentially polished using 6 micron, 3 microns, 0.5 micron, and 0.1-micron diamond pastes.[38, 121] Each step was carried out for 20 minutes in an '8' fashion movement. The final step of polishing was down to 0.05 micron using aluminum slurry. The polished iron piece was ultrasonicated with acetone, thoroughly rinsed with deionized water and ethanol, and dried with nitrogen gas. Finally, the polished, cleaned, and dried piece was stored in the desiccator for a short period before placing it in the sample cell for the SFG experiment. Each step was repeated for the same substrate for the individual SFG experiment.

Sample Preparation

0.4 mM aqueous solution (~5 ml) in D_2O (Cambridge Isotope Laboratories) was prepared for C12 Quat. D_2O solvent was used instead of H_2O solvent to avoid possible CH and OH vibrational mode interferences. A freshly prepared sample was sonicated for ~5 minutes. The solution was then injected into the sample cell containing metal substrate and equilibrated for an hour before collecting SFG spectra.

SFG Experiment

In this chapter, we report the *in-situ* adsorption of Quat and their assembly without any assisting mechanism studied via SFG spectroscopy. The details of SFG theory, sample cell, and SFG experimental setup was discussed in the previous chapters. Gold metal was replaced from the sample cell with the sample of interest iron substrate. The height between the quartz window and the metal substrate adjusted using ~254-micron thickness of Teflon spacer. SFG spectra were collected in CH vibrational region at 2900 cm^{-1}. All the spectra were collected for 3 seconds acquisition for 180 accumulation in total 9 minutes of integration time. Then the spectra were corrected for background, smoothed, and plotted using OriginPro 9.1 software.

Results and Discussion

Preliminary data: Quats self-assembled monolayer formation on Fe substrate

SFG spectrum contains both resonant and nonresonant contributions from the adsorbed molecules and the substrate. The SFG spectra were acquired by overcoming the nonresonant background from the Fe metal substrate. The nonresonant background was suppressed by making a time delay between the visible beam and the IR beam. Time delay minimizes the nonresonant contribution from the substrate and maximizes the resonant contribution from the adsorbed molecules. The time delay was found to be ~2.30 ps and ~2.75 ps for ssp and ppp polarization combinations. The exact delays at Fe substrate were determined experimentally. The SFG spectra were collected at different time delays until the signal from Fe substrate was effectively suppressed (**Figure 6.1** and **Figure 6.2**).

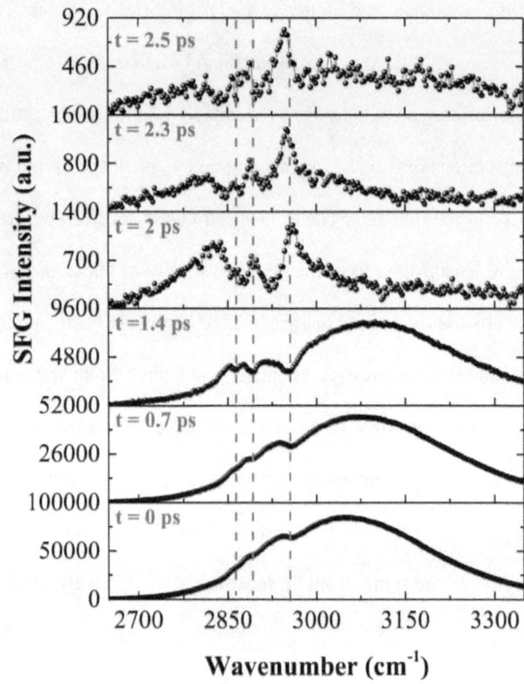

Figure 6.1: SFG spectra of C12 Quat at Fe substrate acquired at different time delays at ssp polarization combinations.

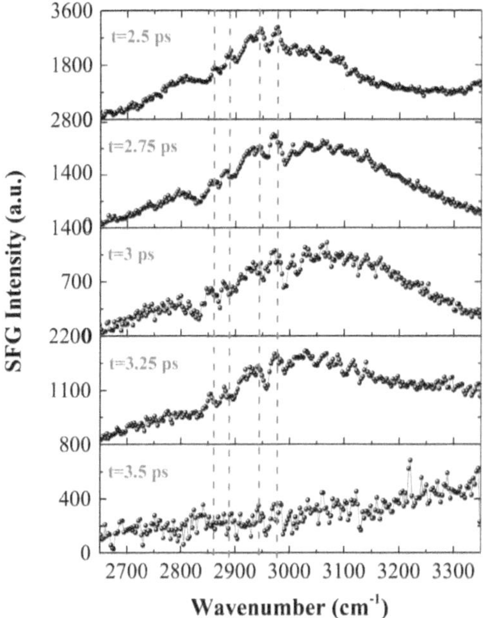

Figure 6.2: SFG spectra of C12 Quat at Fe substrate acquired at different time delays at ppp polarization combinations.

In **Figure 6.3**, the ssp spectrum of C12 Quat showed methylene symmetric stretch peak (CH$_2$ SS) at ~2859 cm^{-1}, terminal methyl symmetric (CH$_3$ SS) at ~2884 cm^{-1}, and the methyl Fermi resonance (CH$_3$ FR); methyl symmetric stretch split by the Fermi resonance interaction with the methyl bending mode at ~2949 cm^{-1}.[57, 131, 132] The peak positions and assignments are listed in **Table 6.1**. The CH$_2$ SS at ~2859 cm^{-1} is barely visible and less prominent compared to the terminal methyl symmetric CH$_3$ SS at ~2884 cm^{-1}. This observation indicates that the Quat molecule adsorbs on Fe substrate and the monolayer contains a small amount of gauche defects.

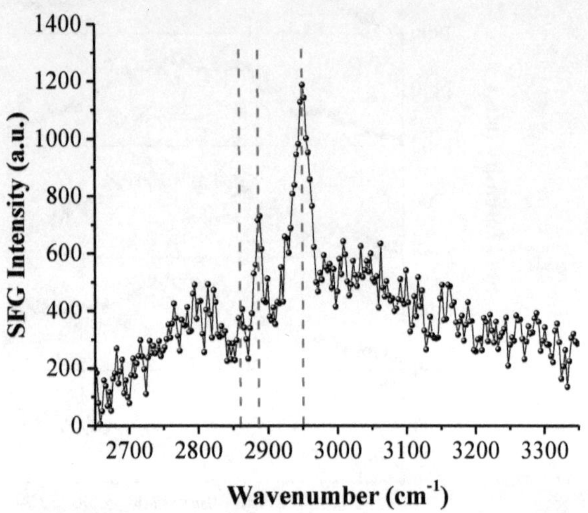

Figure 6.3: SFG spectra of Quat at Fe substrate at ssp polarization combinations.

Table 6.1: Peak assignments of C12 Quat at ssp and ppp polarization combinations.

Peak assignments	ssp wavenumber cm^{-1}	ppp wavenumber cm^{-1}
Methylene symmetric stretch (CH$_2$ SS)[49, 50, 77, 78]	~2859	~2857
Methyl symmetric stretch (CH$_3$ SS)[49, 50, 77, 78]	~2884	~2883
Methyl symmetric stretch split by the Fermi resonance interaction with the methyl bending mode (CH$_3$ FR)[78, 79]	~2949	~2943
Methyl asymmetric stretch (CH$_3$ AS)[133]	-	~2972

In **Figure 6.4**, the ppp spectrum shows the methylene symmetric stretch peak (CH$_2$ SS) at ~2857 cm^{-1}, terminal methyl symmetric (CH$_3$ SS) at ~2883 cm^{-1}, and methyl Fermi stretch (CH$_3$ FR) at ~2943 cm^{-1}, and terminal methyl asymmetric (CH$_3$ AS) at ~2972 cm^{-1}. The metal substrate contains a very high nonresonant signal at ppp polarization combination and complete suppression of nonresonant contribution was not achieved. Several factors will be considered for future experiments to further improve the suppression of nonresonant contribution. These factors are (1) metal surface condition, such as flatness and smoothness, (2) finest polishing of metal substrate without scratching the surface, (3) minimum amount of solution inside the sample cell, (4) and collecting SFG spectra at smaller time delay to achieve the optimum signal at minimum nonresonant contribution.

In ppp polarization (**Figure 6.4**), the peaks of a terminal methyl group at ~2972 cm^{-1} dominant over the peak of methylene group at ~2857 cm^{-1}. This observation also support that Quat form an ordered monolayer with minimum amount of gauche defect.

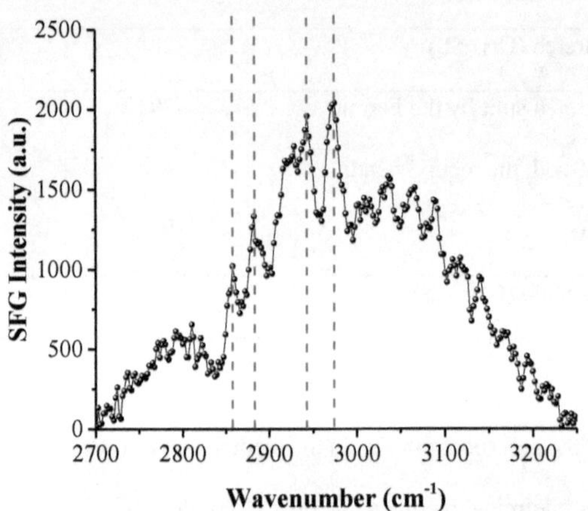

Figure 6.4: SFG spectra of Quat at Fe substrate at ppp polarization combinations.

From the fitted data we can estimate the gauche defect, tilt angle of the terminal methyl group, and distribution of tilt angle to the surface normal. Then, we can compare the orientation of quat monolayer on Fe and Au substrates. This finding leads to our future studies on the effect of metal substrate on the conformational changes of surfactants at liquid-metal interfaces.

CHAPTER 7: SUMMARY AND FUTURE WORK

Summary

Surfactants are known to adsorb on the metal surface and make a barrier or protective thin film against the corrosive substances. Here, we have systematically investigated surfactant adsorption and self-assembled monolayer formation at air-liquid and liquid-metal interfaces by varying their chain length, headgroup, ionic strength, and nature of the metal surface. This research work presented the utilization of surface-specific sum frequency generation (SFG) spectroscopy technique for the molecular-level understanding of conformational changes of surfactants which will help us to design new surfactants to perform at specific conditions.

Chapter 3 discussed the synthesis procedures of five Quaternary ammonium bromide (Quats) and imidazolines. Five Quats were synthesized by the quaternization of amine with respective bromoalkenes. Imidazolines were synthesized from fatty acid and aldehyde at high temperature and room temperature (one-pot synthesis).

Chapter 4 concluded that the Quat series retains surface activity with a significant amount of gauche defects at the air-water interface. The combined effect of increasing chain length and ionic strength dependence was investigated. In general gauche defect increases as a function of increasing chain length was observed. For the shortest chain length of Quat, the gauche defect increases as a function of ionic strength. Increasing alkyl chain length was found to have more ordered interfacial water molecules. However, C4 showed an unexpected behavior due to the orientational transition occurs at the air-water interface. The SFG intensity of the interfacial water molecules was found to have decreased with increasing salt concentration. The headgroup of surfactants affects the orientation of interfacial water molecules. The SFG spectrum of C4 Quat showed more ordered water molecules at the air-liquid interface in comparison to C4 TAB surfactants. Surface tension measurements were also conducted for the comparison of SFG results.

In Chapter 5, we utilized *in-situ* SFG spectroscopy; a powerful tool to investigate the adsorption behavior of Quats onto the gold metal-liquid interface. We studied the self-assembled monolayer formation of cationic surfactants from an aqueous solution and performed orientational analysis. We found to have the longer alkyl tail surfactant formed an ordered monolayer onto the gold surface compared to the shorter alkyl tail surfactant.

Finally, in Chapter 6, the nature of the metal substrate on the conformational changes of surfactants at liquid-metal interfaces was investigated by SFG spectroscopy. The preliminary SFG results on iron (Fe) substrate show that surfactant adsorbs onto Fe surface with the presence of methylene symmetric stretch peak (CH_2 SS) at ~2859 cm^{-1}. This indicates the self-assembled monolayer on the Fe surface is less ordered compared to the self-assembled monolayer on the Au surface.

Future Work

This research work presented in this dissertation is focused on the fundamental adsorption behavior of surfactants at air-water and water-metal interfaces. The results and findings lead to our future work as a next step or the continuation of the project. The following section provides some of the preliminary results in one of the interesting directions on the adsorption and self-assembly of surfactants on stainless steel and mild steel substrate to mimic the actual corrosive environment inside the pipelines are presented below.

Self-assembly of Quat at Steel-D₂O Interface

We have successfully demonstrated the *in-situ* adsorption of Quat and their assembly at gold and iron substrate without potential studied via SFG spectroscopy. Here we provide preliminary results on mild steel and stainless steel substrate. The mild steel and stainless steel substrate was polished down to 0.05micron using diamond paste and aluminum slurry. Gold metal in the sample cell was replaced by mild steel and stainless steel, respectively. The height between the quartz window and the metal substrate adjusted using ~254-micron thickness of Teflon spacer. SFG spectra were collected in CH vibrational region at 2900 cm^{-1}.

In **Figure 7.1**, the ssp spectrum shows the terminal methyl symmetric (CH₃ SS) at ~2882 cm^{-1} and methyl Fermi stretch (CH₃ FR) at ~2947 cm^{-1}. The CH₂ SS at ~2859 cm^{-1} is barely visible due to the spectral convolution and less prominent compared to the terminal methyl symmetric CH₃ SS at ~2884 cm^{-1}.

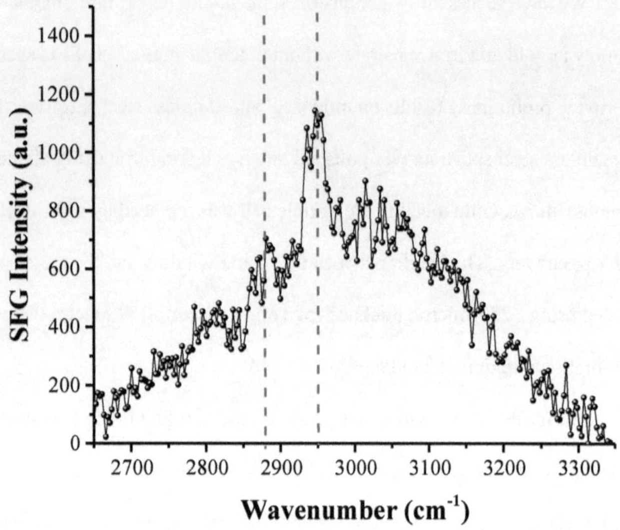

Figure 7.1: SFG spectra of Quat at mild steel substrate at ssp polarization combinations.

This preliminary data indicates that the Quat molecule adsorbs and forms a self-assembled monolayer on mild steel surface. The monolayer contains small amount of gauche defects.

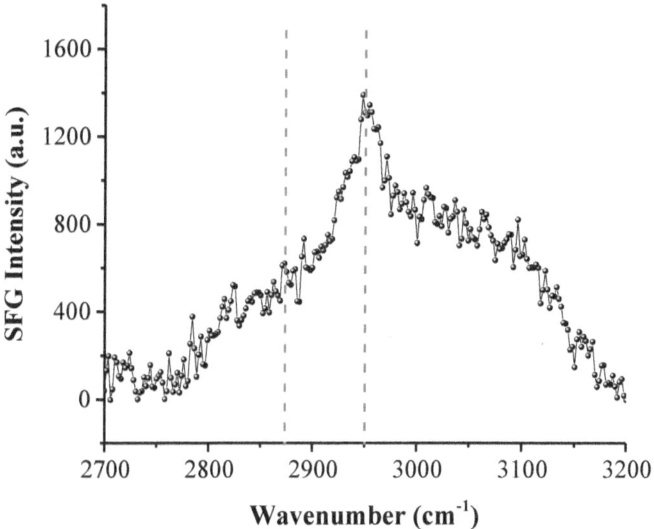

Figure 7.2: SFG spectra of Quat at stainless steel substrate at ssp polarization combinations.

In **Figure 7.2,** the ssp spectrum shows the methyl Fermi stretch (CH_3 FR) at ~2949 cm^{-1}.[57, 131, 132] The methylene CH_2 SS at ~2859 cm^{-1} and the terminal methyl symmetric (CH_3 SS) at ~2882 cm^{-1} are barely visible due to the spectral convolution. This preliminary data indicates that the Quat molecule adsorbs and forms a self-assembled monolayer (SAM) on the stainless steel surface. This preliminary result is important as it set the platform for providing the molecular-level understanding of SAM formation on oil and gas pipelines compared to the noble metals.

However, a better signal-to-noise ratio and nonresonant background suppression are needed for the orientational analysis. Further experiments will be required at mild steel and stainless steel substrate. From the continuation of this project, we will be able to estimate the gauche defect, the average tilt angle of the terminal methyl group, and the distribution of tilt angle to the surface normal. Then, we will be able to compare the effect of substrate on conformational changes of quat monolayer at liquid-metal interfaces. The extension of this project by mimicking a corrosive environment will lead to the design of a better surfactant for the corrosion inhibition process. This is an ongoing project in collaboration with the Institute for Corrosion and Multiphase Technology (ICMT), the Department of Chemistry and Biochemistry, and the Department of Molecular and Biomolecular Engineering at Ohio University. This project includes various techniques, such as electrochemical quartz crystal microbalance, electrochemical impedance spectroscopy, fluorometry, and theoretical studies. Ultimately, we can extract more detailed information on surfactant's adsorption and self-assembled formation on air-liquid and liquid-solid interfaces.

REFERENCES

1. Koch, G. H. *Historic congresiional study: Corrosion cost and preventive strategies in the United States*; **2002**.

2. Koch, G. H.; Brongers, M. P. H.; Thompson, N. G.; Virmani, Y. P.; Payer, J. H. *Corrosion cost and preventive strategies in the United States*; **2002**.

3. Rhodes, P. R., Environment-assisted cracking of corrosion-resistant alloys in oil and gas production environments: a review. *Corrosion* **2001**, *57* (11), 923-966.

4. Kendig, M. W.; Buchheit, R. G., Corrosion inhibition of aluminum and aluminum alloys by soluble chromates, chromate coatings, and chromate-free coatings. *Corrosion* **2003**, *59* (5), 379-400.

5. Shipilov, S. A.; Le May, I., Structural integrity of aging buried pipelines having cathodic protection. *Engineering Failure Analysis* **2006**, *13* (7), 1159-1176.

6. McMahon, A. J., The mechanism of action of an oleic imidazoline based corrosion inhibitor for oilfield use. *Colloids and Surfaces* **1991**, *59*, 187-208.

7. Finšgar, M.; Jackson, J., Application of corrosion inhibitors for steels in acidic media for the oil and gas industry: a review. *Corrosion Science* **2014**, *86*, 17-41.

8. Popoola, L. T.; Grema, A. S.; Latinwo, G. K.; Gutti, B.; Balogun, A. S., Corrosion problems during oil and gas production and its mitigation. *International Journal of Industrial Chemistry* **2013**, *4* (1), 35.

9. Nešić, S., Key issues related to modelling of internal corrosion of oil and gas pipelines–A review. *Corrosion science* **2007**, *49* (12), 4308-4338.

10. Kermani, M. B.; Morshed, A., Carbon dioxide corrosion in oil and gas production—a compendium. *Corrosion* **2003**, *59* (8), 659-683.

11. Nešić, S., Carbon dioxide corrosion of mild steel. *Uhlig's Corrosion Handbook* **2011**, 229-245.

12. Tran, T.; Brown, B.; Nesic, S., Corrosion of mild steel in an aqueous CO2 environment—basic electrochemical mechanisms revisited. *Corrosion* **2015**.

13. De Waard, C.; Milliams, D. E., Carbonic acid corrosion of steel. *Corrosion* **1975**, *31* (5), 177-181.

14. Li, C.; Richter, S.; Nešićl, S., How do inhibitors mitigate corrosion in oil-water two-phase flow beyond lowering the corrosion rate? *Corrosion* **2014**, *70* (9), 958-966.

15. Tang, Y.; Yang, X.; Yang, W.; Wan, R.; Chen, Y.; Yin, X., A preliminary investigation of corrosion inhibition of mild steel in 0.5 M H2SO4 by 2-amino-5-(n-pyridyl)-1, 3, 4-thiadiazole: polarization, EIS and molecular dynamics simulations. *Corrosion Science* **2010**, *52* (5), 1801-1808.

16. Tan, Y.-J.; Bailey, S.; Kinsella, B., An investigation of the formation and destruction of corrosion inhibitor films using electrochemical impedance spectroscopy (EIS). *Corrosion science* **1996**, *38* (9), 1545-1561.

17.   Liu, X.; Chen, S.; Tian, F.; Ma, H.; Shen, L.; Zhai, H., Studies of protection of iron corrosion by rosin imidazoline self-assembled monolayers. *Surface and interface analysis* **2007**, *39* (4), 317-323.

18.   Edwards, A.; Osborne, C.; Webster, S.; Klenerman, D.; Joseph, M.; Ostovar, P.; Doyle, M., Mechanistic studies of the corrosion inhibitor oleic imidazoline. *Corrosion Science* **1994**, *36* (2), 315-325.

19.   Xiong, Y.; Brown, B.; Kinsella, B.; Nešić, S.; Pailleret, A., Atomic force microscopy study of the adsorption of surfactant corrosion inhibitor films. *Corrosion* **2013**, *70* (3), 247-260.

20.   Cruz, J.; Martınez, R.; Genesca, J.; Garcıa-Ochoa, E., Experimental and theoretical study of 1-(2-ethylamino)-2-methylimidazoline as an inhibitor of carbon steel corrosion in acid media. *Journal of Electroanalytical Chemistry* **2004**, *566* (1), 111-121.

21.   Nesic, S.; Wilhelmsen, W.; Skjerve, S.; Hesjevik, S. M. In *Testing of inhibitors for CO2 corrosion using the electrochemical techniques*, 1995; p 1163.

22.   Badr, E. A., Inhibition effect of synthesized cationic surfactant on the corrosion of carbon steel in 1 M HCl. *Journal of Industrial and Engineering Chemistry* **2014**, *20* (5), 3361-3366.

23.   Ramachandran, S.; Tsai, B.-L.; Blanco, M.; Chen, H.; Tang, Y.; Goddard, W. A., Self-assembled monolayer mechanism for corrosion inhibition of iron by imidazolines. *Langmuir* **1996**, *12* (26), 6419-6428.

24.   Cimatu, K. A.; Chan, S. C.; Jang, J. H.; Hafer, K., Preferential organization of methacrylate monomers and polymer thin films at the air interface using femtosecond sum frequency generation spectroscopy. *The Journal of Physical Chemistry C* **2015**, *119* (45), 25327-25339.

25.   Lambert, A. G.; Davies, P. B.; Neivandt, D. J., Implementing the theory of sum frequency generation vibrational spectroscopy: A tutorial review. *Applied Spectroscopy Reviews* **2005**, *40* (2), 103-145.

26.   Kim, J.; Somorjai, G. A., Molecular packing of lysozyme, fibrinogen, and bovine serum albumin on hydrophilic and hydrophobic surfaces studied by infrared–visible sum frequency generation and fluorescence microscopy. *Journal of the American Chemical Society* **2003**, *125* (10), 3150-3158.

27.   Maroncelli, M.; Strauss, H. L.; Snyder, R. G., The distribution of conformational disorder in the high-temperature phases of the crystalline n-alkanes. *The Journal of chemical physics* **1985**, *82* (6), 2811-2824.

28.   Cimatu, K. A.; Chan, S. C.; Jang, J. H.; Hafer, K., Preferential Organization of Methacrylate Monomers and Polymer Thin Films at the Air Interface Using Femtosecond Sum Frequency Generation Spectroscopy. *J Phys Chem C* **2015**, *119* (45), 25327-25339.

29.   Chan, S. C.; Jang, J. H.; Cimatu, K. A., Orientational Analysis of Interfacial Molecular Groups of a 2-Methoxyethyl Methacrylate Monomer Using

Femtosecond Sum Frequency Generation Spectroscopy. *The Journal of Physical Chemistry C* **2016,** *120* (51), 29358-29373.

30.   Kermani, M. B.; Harrop, D., The impact of corrosion on oil and gas industry. *SPE Production & Facilities* **1996,** *11* (03), 186-190.

31.   Roberge, P. R.; Eng, P., Corrosion Engineering. *Principles and Practice* **2005,** *1.*

32.   Nmai, C. K.; Farrington, S. A.; Bobroski, G., Organic-based corrosion-inhibiting admixture for reinforced concrete. *Concrete International* **1992,** *14* (4), 45-51.

33.   Jevremović, I.; Singer, M.; Nešić, S.; Mišković-Stanković, V., Inhibition properties of self-assembled corrosion inhibitor talloil diethylenetriamine imidazoline for mild steel corrosion in chloride solution saturated with carbon dioxide. *Corrosion Science* **2013,** *77,* 265-272.

34.   Knag, M.; Sjöblom, J.; Øye, G.; Gulbrandsen, E., A quartz crystal microbalance study of the adsorption of quaternary ammonium derivates on iron and cementite. *Colloids and Surfaces A: Physicochemical and Engineering Aspects* **2004,** *250* (1-3), 269-278.

35.   Vračar, L. M.; Dražić, D. M., Adsorption and corrosion inhibitive properties of some organic molecules on iron electrode in sulfuric acid. *Corrosion Science* **2002,** *44* (8), 1669-1680.

36.   Cao, C., On electrochemical techniques for interface inhibitor research. *corrosion science* **1996,** *38* (12), 2073-2082.

37.   Cimatu, K.; Baldelli, S., Sum frequency generation microscopy of microcontact-printed mixed self-assembled monolayers. *The Journal of Physical Chemistry B* **2006,** *110* (4), 1807-1813.

38.   Cimatu, K.; Baldelli, S., Spatially resolved surface analysis of an octadecanethiol self-assembled monolayer on mild steel using sum frequency generation imaging microscopy. *The Journal of Physical Chemistry C* **2007,** *111* (19), 7137-7143.

39.   Cimatu, K.; Moore, H. J.; Barriet, D.; Chinwangso, P.; Lee, T. R.; Baldelli, S., Sum Frequency Generation Imaging Microscopy of Patterned Self-Assembled Monolayers with Terminal− CH3,− OCH3,− CF2CF3,− C☐ C,− Phenyl, and− Cyclopropyl Groups. *The Journal of Physical Chemistry C* **2008,** *112* (37), 14529-14537.

40.   Pizzolatto, R. L.; Yang, Y. J.; Wolf, L. K.; Messmer, M. C., Conformational aspects of model chromatographic surfaces studied by sum-frequency generation. *Analytica chimica acta* **1999,** *397* (1-3), 81-92.

41.   Ebert, V.; Schulz, C.; Volpp, H. R.; Wolfrum, J.; Monkhouse, P., Laser diagnostics of combustion processes: from chemical dynamics to technical devices. *Israel journal of chemistry* **1999,** *39* (1), 1-24.

42.   Beattie, D. A.; Haydock, S.; Bain, C. D., A comparative study of confined organic monolayers by Raman scattering and sum-frequency spectroscopy. *Vibrational Spectroscopy* **2000,** *24* (1), 109-123.

43. Bell, G. R.; Bain, C. D.; Ward, R. N., Sum-frequency vibrational spectroscopy of soluble surfactants at the air/water interface. *Journal of the Chemical Society, Faraday Transactions* **1996,** *92* (4), 515-523.

44. Braun, R.; Casson, B. D.; Bain, C. D., A sum-frequency study of the two-dimensional phase transition in a monolayer of undecanol on water. *Chemical physics letters* **1995,** *245* (4-5), 326-334.

45. Ward, R. N.; Duffy, D. C.; Davies, P. B.; Bain, C. D., Sum-frequency spectroscopy of surfactants adsorbed at a flat hydrophobic surface. *The Journal of Physical Chemistry* **1994,** *98* (34), 8536-8542.

46. Ward, R. N.; Davies, P. B.; Bain, C. D., Coadsorption of sodium dodecyl sulfate and dodecanol at a hydrophobic surface. *The Journal of Physical Chemistry B* **1997,** *101* (9), 1594-1601.

47. Sommer, H. Z.; Lipp, H. I.; Jackson, L. L., Alkylation of amines. General exhaustive alkylation method for the synthesis of quaternary ammonium compounds. *The Journal of Organic Chemistry* **1971,** *36* (6), 824-828.

48. Patai, S., chemistry of the amino group. **1968.**

49. Khan, M. R.; Premadasa, U. I.; Cimatu, K. L. A., Role of the Cationic Headgroup to Conformational Changes Undergone by Shorter Alkyl Chain Surfactant and Water Molecules at the Air-Liquid Interface. *Journal of Colloid and Interface Science* **2020.**

50. Premadasa, U. I.; Moradighadi, N.; Kotturi, K.; Nonkumwong, J.; Khan, M. R.; Singer, M.; Masson, E.; Cimatu, K. L. A., Solvent Isotopic Effects on a Surfactant Headgroup at the Air–Liquid Interface. *The Journal of Physical Chemistry C* **2018,** *122* (28), 16079-16085.

51. Pracht, H. J.; Nirschl, J. P., Process for making imidazolinium salts, fabric conditioning compositions and methods. Google Patents: **1978.**

52. Bajpai, D.; Tyagi, V. K., Fatty imidazolines: chemistry, synthesis, properties and their industrial applications. *Journal of oleo science* **2006,** *55* (7), 319-329.

53. Gill, J. S.; Reed, P. E.; Banerjee, S.; Harbindu, A., Water soluble substituted imidazolines as corrosion inhibitors for ferrous metals. Google Patents: **2017.**

54. Martin, J. A.; Valone, F. W., The existence of imidazoline corrosion inhibitors. *Corrosion* **1985,** *41* (5), 281-287.

55. Fujioka, H.; Murai, K.; Kubo, O.; Ohba, Y.; Kita, Y., One-pot synthesis of imidazolines from aldehydes: detailed study about solvents and substrates. *Tetrahedron* **2007,** *63* (3), 638-643.

56. Hao, D.; Yun-lei, Z.; Xiao-peng, S.; Jin-ming, Y.; Dong, F., Synthesis of 2-amino-4-phenyl-6-(phenylsulfanyl)-3, 5-dicyanopyridines by tandem reaction. *Research on Chemical Intermediates* **2014,** *40* (2), 587-594.

57. Khan, M. R.; Singh, H.; Sharma, S.; Asetre Cimatu, K. L., Direct Observation of Adsorption Morphologies of Cationic Surfactants at the Gold Metal–Liquid Interface. *The Journal of Physical Chemistry Letters* **2020**, *11* (22), 9901-9906.

58. Migahed, M. A.; Al-Sabagh, A. M., Beneficial role of surfactants as corrosion inhibitors in petroleum industry: a review article. *Chemical Engineering Communications* **2009**, *196* (9), 1054-1075.

59. Zhu, Y.; FreeM, M. L., Effects of surfactant aggregation and adsorption on steel corrosion inhibition in salt solution. *Polymer Sciences* **2015**.

60. Gragson, D. E.; Richmond, G. L., Probing the structure of water molecules at an oil/water interface in the presence of a charged soluble surfactant through isotopic dilution studies. *The Journal of Physical Chemistry B* **1998**, *102* (3), 569-576.

61. Lin, I. J.; Moudgil, B. M.; Somasundaran, P., Estimation of the effective number of—CH 2-groups in long-chain surface active agents. *Colloid and Polymer Science* **1974**, *252* (5), 407-414.

62. Sung, W.; Avazbaeva, Z.; Kim, D., Salt promotes protonation of amine groups at air/water interface. *The journal of physical chemistry letters* **2017**, *8* (15), 3601-3606.

63. Gragson, D. E.; McCarty, B. M.; Richmond, G. L., Surfactant/water interactions at the air/water interface probed by vibrational sum frequency generation. *The Journal of Physical Chemistry* **1996**, *100* (34), 14272-14275.

64. Aliaga, C.; Baker, G. A.; Baldelli, S., Sum frequency generation studies of ammonium and pyrrolidinium ionic liquids based on the bis-trifluoromethanesulfonimide anion. *The Journal of Physical Chemistry B* **2008**, *112* (6), 1676-1684.

65. Jena, K. C.; Covert, P. A.; Hore, D. K., The effect of salt on the water structure at a charged solid surface: Differentiating second-and third-order nonlinear contributions. *The Journal of Physical Chemistry Letters* **2011**, *2* (9), 1056-1061.

66. Shahir, A. A.; Khristov, K.; Nguyen, K. T.; Nguyen, A. V.; Mileva, E., Combined sum frequency generation and thin liquid film study of the specific effect of monovalent cations on the interfacial water structure. *Langmuir* **2018**, *34* (23), 6844-6855.

67. Reitböck, C.; Głowacki, E.; Stifter, D., Sum-Frequency Generation Vibrational Spectroscopy Investigations of Phosphonic Acids on Anodic Aluminum Oxide Films. *Applied spectroscopy* **2018**, *72* (5), 725-730.

68. Conboy, J. C.; Messmer, M. C.; Richmond, G. L., Investigation of surfactant conformation and order at the liquid– liquid interface by total internal reflection sum-frequency vibrational spectroscopy. *The Journal of Physical Chemistry* **1996**, *100* (18), 7617-7622.

69. Conboy, J. C.; Messmer, M. C.; Richmond, G. L., Effect of alkyl chain length on the conformation and order of simple ionic surfactants adsorbed at the D2O/CCl4

interface as studied by sum-frequency vibrational spectroscopy. *Langmuir* **1998,** *14* (23), 6722-6727.

70.   Liu, D.; Ma, G.; Levering, L. M.; Allen, H. C., Vibrational spectroscopy of aqueous sodium halide solutions and air– liquid interfaces: Observation of increased interfacial depth. *The Journal of Physical Chemistry B* **2004,** *108* (7), 2252-2260.

71.   Iimori, T.; Iwahashi, T.; Kanai, K.; Seki, K.; Sung, J.; Kim, D.; Hamaguchi, H.-o.; Ouchi, Y., Local Structure at the Air/Liquid Interface of Room-Temperature Ionic Liquids Probed by Infrared– Visible Sum Frequency Generation Vibrational Spectroscopy: 1-Alkyl-3-methylimidazolium Tetrafluoroborates. *The Journal of Physical Chemistry B* **2007,** *111* (18), 4860-4866.

72.   Dogangun, M.; Ohno, P. E.; Liang, D.; McGeachy, A. C.; Be, A. G.; Dalchand, N.; Li, T.; Cui, Q.; Geiger, F. M., Hydrogen-Bond Networks near Supported Lipid Bilayers from Vibrational Sum Frequency Generation Experiments and Atomistic Simulations. *The Journal of Physical Chemistry B* **2018,** *122* (18), 4870-4879.

73.   Jena, K. C.; Hore, D. K., Variation of ionic strength reveals the interfacial water structure at a charged mineral surface. *The Journal of Physical Chemistry C* **2009,** *113* (34), 15364-15372.

74.   Nihonyanagi, S.; Yamaguchi, S.; Tahara, T., Counterion effect on interfacial water at charged interfaces and its relevance to the Hofmeister series. *Journal of the American Chemical Society* **2014,** *136* (17), 6155-6158.

75.   Wang*, H.-F.; Gan, W.; Lu†‡ §, R.; Rao†‡¶, Y.; Wu, B.-H., Quantitative spectral and orientational analysis in surface sum frequency generation vibrational spectroscopy (SFG-VS). *International Reviews in Physical Chemistry* **2005,** *24* (2), 191-256.

76.   Duplan, J. C.; Mahi, L.; Brunet, J. L., NMR determination of the equilibrium constant for the liquid H2O–D2O mixture. *Chemical physics letters* **2005,** *413* (4-6), 400-403.

77.   Quast, A. D.; Wilde, N. C.; Matthews, S. S.; Maughan, S. T.; Castle, S. L.; Patterson, J. E., Improved assignment of vibrational modes in sum-frequency spectra in the CH stretch region for surface-bound C18 alkylsilanes. *Vibrational Spectroscopy* **2012,** *61*, 17-24.

78.   Richter, L. J.; Petralli-Mallow, T. P.; Stephenson, J. C., Vibrationally resolved sum-frequency generation with broad-bandwidth infrared pulses. *Optics Letters* **1998,** *23* (20), 1594-1596.

79.   Verreault, D.; Kurz, V.; Howell, C.; Koelsch, P., Sample cells for probing solid/liquid interfaces with broadband sum-frequency-generation spectroscopy. *Review of Scientific Instruments* **2010,** *81* (6), 063111.

80.   Wiley, J., *Sons*. Inc New York; NY: **2001.**

81.	Tyrode, E.; Rutland, M. W.; Bain, C. D., Adsorption of CTAB on hydrophilic silica studied by linear and nonlinear optical spectroscopy. *Journal of the American Chemical Society* **2008**, *130* (51), 17434-17445.

82.	Nguyen, K. T.; Nguyen, A. V., Suppressing interfacial water signals to assist the peak assignment of the N+–H stretching mode in sum frequency generation vibrational spectroscopy. *Physical Chemistry Chemical Physics* **2015**, *17* (43), 28534-28538.

83.	Dutta, C.; Svirida, A.; Mammetkuliyev, M.; Rukhadze, M.; Benderskii, A. V., Insight into Water Structure at the Surfactant Surfaces and in Microemulsion Confinement. *The Journal of Physical Chemistry B* **2017**, *121* (31), 7447-7454.

84.	Argyris, D.; Cole, D. R.; Striolo, A., Ion-specific effects under confinement: the role of interfacial water. *Acs Nano* **2010**, *4* (4), 2035-2042.

85.	Nihonyanagi, S.; Yamaguchi, S.; Tahara, T., Direct evidence for orientational flip-flop of water molecules at charged interfaces: A heterodyne-detected vibrational sum frequency generation study. *The Journal of chemical physics* **2009**, *130* (20), 204704.

86.	Du, Q.; Superfine, R.; Freysz, E.; Shen, Y. R., Vibrational spectroscopy of water at the vapor/water interface. *Physical Review Letters* **1993**, *70* (15), 2313.

87.	Bain, C. D., Studies of adsorption at interfaces by optical techniques: ellipsometry, second harmonic generation and sum-frequency generation. *Current opinion in colloid & interface science* **1998**, *3* (3), 287-292.

88.	Nguyen, K. T.; Nguyen, A. V.; Evans, G. M., Interfacial Water Structure at Surfactant Concentrations below and above the Critical Micelle Concentration as Revealed by Sum Frequency Generation Vibrational Spectroscopy. *The Journal of Physical Chemistry C* **2015**, *119* (27), 15477-15481.

89.	Feng, R.-J.; Li, X.; Zhang, Z.; Lu, Z.; Guo, Y., Spectral assignment and orientational analysis in a vibrational sum frequency generation study of DPPC monolayers at the air/water interface. *The Journal of chemical physics* **2016**, *145* (24), 244707.

90.	Ishiyama, T.; Sokolov, V. V.; Morita, A., Molecular dynamics simulation of liquid methanol. II. Unified assignment of infrared, raman, and sum frequency generation vibrational spectra in methyl C–H stretching region. *The Journal of chemical physics* **2011**, *134* (2), 024510.

91.	Bzeih, W.; Gheribi, A.; Wood-Adams, P. M.; Hayes, P. L., Dependence of the Surface Structure of Polystyrene on Chain Molecular Weight Investigated by Sum Frequency Generation Spectroscopy. *The Journal of Physical Chemistry C* **2018**.

92.	Goussous, S. A.; Casford, M. T. L.; Johnson, S. A.; Davies, P. B., A structural and temporal study of the surfactants behenyltrimethylammonium methosulfate and behenyltrimethylammonium chloride adsorbed at air/water and air/glass interfaces using sum frequency generation spectroscopy. *Journal of colloid and interface science* **2017**, *488*, 365-372.

93.   Knock, M. M.; Bain, C. D., Effect of counterion on monolayers of hexadecyltrimethylammonium halides at the air– water interface. *Langmuir* **2000,** *16* (6), 2857-2865.

94.   Dederichs, F.; Friedrich, K. A.; Daum, W., Sum-frequency vibrational spectroscopy of CO adsorption on Pt (111) and Pt (110) electrode surfaces in perchloric acid solution: effects of thin-layer electrolytes in spectroelectrochemistry. *The Journal of Physical Chemistry B* **2000,** *104* (28), 6626-6632.

95.   Day, B. S.; Shuler, S. F.; Ducre, A.; Morris, J. R., The dynamics of gas-surface energy exchange in collisions of Ar atoms with ω-functionalized self-assembled monolayers. *The Journal of chemical physics* **2003,** *119* (15), 8084-8096.

96.   Premadasa, U. I.; Adhikari, N. M.; Baral, S.; Aboelenen, A. M.; Cimatu, K. L. A., Conformational Changes of Methacrylate-Based Monomers at the Air–Liquid Interface Due to Bulky Substituents. *The Journal of Physical Chemistry C* **2017,** *121* (31), 16888-16902.

97.   Shahir, A. A.; Nguyen, K. T.; Nguyen, A. V., A sum-frequency generation spectroscopic study of the Gibbs analysis paradox: monolayer or sub-monolayer adsorption? *Physical Chemistry Chemical Physics* **2016,** *18* (13), 8794-8805.

98.   Rao, Y.; Li, X.; Lei, X.; Jockusch, S.; George, M. W.; Turro, N. J.; Eisenthal, K. B., Observations of interfacial population and organization of surfactants with sum frequency generation and surface tension. *The Journal of Physical Chemistry C* **2011,** *115* (24), 12064-12067.

99.   Menger, F. M.; Rizvi, S. A. A., Relationship between surface tension and surface coverage. *Langmuir* **2011,** *27* (23), 13975-13977.

100.   Butt, H. J.; Graf, K.; Kappl, M., Thermodynamics of interfaces. *Physics and Chemistry of Interfaces, 3rd ed.; Wiley-VCH: Weinheim, Germany* **2004,** 26-41.

101.   Mondal, J. A.; Namboodiri, V.; Mathi, P.; Singh, A. K., Alkyl chain length dependent structural and orientational transformations of water at alcohol–water interfaces and its relevance to atmospheric aerosols. *The journal of physical chemistry letters* **2017,** *8* (7), 1637-1644.

102.   Gragson, D. E.; McCarty, B. M.; Richmond, G. L., Ordering of interfacial water molecules at the charged air/water interface observed by vibrational sum frequency generation. *Journal of the American Chemical Society* **1997,** *119* (26), 6144-6152.

103.   Bielawska, M.; Zdziennicka, A.; Jańczuk, B., Comparison between surface and volumetric properties of short-chain alcohols and some classical surfactants. *Annales Universitatis Mariae Curie-Sklodowska, sectio AA–Chemia* **2016,** *71* (1).

104.   Long, J. A.; Rankin, B. M.; Ben-Amotz, D., Micelle structure and hydrophobic hydration. *Journal of the American Chemical Society* **2015,** *137* (33), 10809-10815.

105. Askari, M.; Aliofkhazraei, M.; Ghaffari, S.; Hajizadeh, A., Film former corrosion inhibitors for oil and gas pipelines-A technical review. *Journal of Natural Gas Science and Engineering* **2018**, *58*, 92-114.

106. Yu, D.; Huang, X.; Deng, M.; Lin, Y.; Jiang, L.; Huang, J.; Wang, Y., Effects of inorganic and organic salts on aggregation behavior of cationic gemini surfactants. *The Journal of Physical Chemistry B* **2010**, *114* (46), 14955-14964.

107. Bothorel, P.; Belle, J.; Lemaire, B., Theoretical study of aliphatic chain structure in mono-and bilayers. *Chemistry and physics of lipids* **1974**, *12* (2), 96-116.

108. Fuchs-Godec, R., The adsorption, CMC determination and corrosion inhibition of some N-alkyl quaternary ammonium salts on carbon steel surface in 2 M H2SO4. *Colloids and Surfaces A: Physicochemical and Engineering Aspects* **2006**, *280* (1-3), 130-139.

109. McMahon, A. J., The Mechanism of Action of an Oleic Imidazoline based Corrosion Inhibitor for Oilfield Use. *Colloid Surface* **1991**, *59*, 187-208.

110. Popoola, L. T.; Grema, A. S.; Latinwo, G. K.; Gutti, B.; Balogun, A. S., Corrosion Problems During Oil and Gas Production and its Mitigation. *Int. J. Ind. Chem.* **2013**, *4* (1), 1-15.

111. Kermani, M. B.; Harr, D. In *The impact of corrosion on oil and gas industry*, Giornata di studio IGF S. Donato Milanese **1996**, **2008**.

112. Edwards, A.; Osborne, C.; Webster, S.; Klenerman, D.; Joseph, M.; Ostovar, P.; Doyle, M., Mechanistic Studies of the Corrosion Inhibitor Oleic Imidazoline. *Corros. Sci.* **1994**, *36* (2), 315-325.

113. Somasundaran, P.; Fuerstenau, D. W., Mechanisms of alkyl sulfonate adsorption at the alumina-water interface. *J. Phys. Chem.* **1966**, *70* (1), 90-96.

114. Fan, A.; Somasundaran, P.; Turro, N. J., Adsorption of Alkyltrimethylammonium Bromides on Negatively Charged Alumina. *Langmuir* **1997**, *13* (3), 506-510.

115. Jaschke, M.; Butt, H.-J.; Gaub, H. E.; Manne, S., Surfactant Aggregates at a Metal Surface. *Langmuir* **1997**, *13* (6), 1381-1384.

116. Orendorff, C. J.; Gole, A.; Sau, T. K.; Murphy, C. J., Surface-enhanced Raman spectroscopy of self-assembled monolayers: sandwich architecture and nanoparticle shape dependence. *Anal. Chem.* **2005**, *77* (10), 3261-3266.

117. Wood, M. H.; Welbourn, R. J. L.; Charlton, T.; Zarbakhsh, A.; Casford, M. T.; Clarke, S. M., Hexadecylamine Adsorption at the Iron Oxide–Oil Interface. *Langmuir* **2013**, *29* (45), 13735-13742.

118. Schultz, Z. D.; Biggin, M. E.; White, J. O.; Gewirth, A. A., Infrared−Visible Sum Frequency Generation Investigation of Cu Corrosion Inhibition with Benzotriazole. *Anal Chem* **2004**, *76* (3), 604-609.

119. Miranda, P. B.; Pflumio, V.; Saijo, H.; Shen, Y. R., Surfactant monolayers at solid-liquid interfaces: conformation and interaction. *Thin Solid Films* **1998**, *327*, 161-165.

120. Harris, A. L.; Chidsey, C. E. D.; Levinos, N. J.; Loiacono, D. N., Monolayer vibrational spectroscopy by infrared-visible sum generation at metal and semiconductor surfaces. *Chem Phys Lett* **1987,** *141* (4), 350-356.

121. Zhang, H. P.; Romero, C.; Baldelli, S., Preparation of Akanethiol Monolayers on Mild Steel Surfaces Studied with Sum Frequency Generation and Eletrochemistry. *J. Phys. Chem. B* **2005,** *109* (15520-15530).

122. Shi, H.; Cai, Z.; Patrow, J.; Zhao, B.; Wang, Y.; Wang, Y.; Benderskii, A.; Dawlaty, J.; Cronin, S. B., Monitoring Local Electric Fields at Electrode Surfaces Using Surface Enhanced Raman Scattering-Based Stark-Shift Spectroscopy during Hydrogen Evolution Reactions. *Acs Appl Mater Inter* **2018,** *10* (39), 33678-33683.

123. Ye, S.; Nihonyanagi, S.; Fujishima, K.; Uosaki, K., Conformational Order of Octadecanethiol (ODT) Monolayer at Gold/ Solution Interface: Internal Reflection Sum Frequency Generation (SFG) Study. *Studies in Surface Science and Catalysis* **2001,** *132,* 705-710.

124. Cimatu, K. A.; Baldelli, S., Chemical Imaging of Corrosion: Sum Frequency Generation Imaging Microscopy of Gold Surface in Cyanide. *J. Am. Chem. Soc* **2008,** *130* (25), 8030-8037.

125. Wallentine, S.; Bandaranayake, S.; Biswas, S.; Baker, L. R., Plasmon-Resonant Vibrational Sum Frequency Generation of Electrochemical Interfaces: Direct Observation of Carbon Dioxide Electroreduction on Gold. *The Journal of Physical Chemistry A* **2020,** *124* (39), 8057-8064.

126. Cimatu, K.; Baldelli, S., Sum frequency generation microscopy of microcontact-printed mixed self-assembled monolayers. *J Phys Chem B* **2006,** *110* (4), 1807-13.

127. Nonkumwong, J.; Erasquin, U. J.; Sy Piecco, K. W.; Premadasa, U. I.; Aboelenen, A. M.; Tangonan, A.; Chen, J.; Ingram, D.; Srisombat, L.; Cimatu, K. L. A., Successive Surface Reactions on Hydrophilic Silica for Modified Magnetic Nanoparticle Attachment Probed by Sum-Frequency Generation Spectroscopy. *Langmuir* **2018,** *34* (43), 12680-12693.

128. Adhikari, N. M.; Premadasa, U. I.; Cimatu, K. L. A., Sum frequency generation vibrational spectroscopy of methacrylate-based functional monomers at the hydrophilic solid-liquid interface. *Phys Chem Chem Phys* **2017**.

129. Livingstone, R. A.; Nagata, Y.; Bonn, M.; Backus, E. H. G., Two types of water at the water–surfactant interface revealed by time-resolved vibrational spectroscopy. *Journal of the American Chemical Society* **2015,** *137* (47), 14912-14919.

130. Adhikari, N. M.; Premadasa, U. I.; Rudy, Z. J.; Cimatu, K. L. A., Orientational Analysis of Monolayers at Low Surface Concentrations Due to an Increased Signal-to-Noise Ratio (S/N) Using Broadband Sum Frequency Generation Vibrational Spectroscopy. *Appl Spectrosc* **2019,** *73* (10), 1146-1159.

131. Wang, W.; Ye, S., Molecular interactions of organic molecules at the air/water interface investigated by sum frequency generation vibrational spectroscopy. *Physical Chemistry Chemical Physics* **2017**, *19* (6), 4488-4493.

132. Nishi, N.; Hobara, D.; Yamamoto, M.; Kakiuchi, T., Total-internal-reflection broad-bandwidth sum frequency generation spectroscopy of hexadecanethiol adsorbed on thin gold film deposited on CaF2. *Analytical sciences* **2003**, *19* (6), 887-890.

133. Ma, G.; Allen, H. C., Surface studies of aqueous methanol solutions by vibrational broad bandwidth sum frequency generation spectroscopy. *The Journal of Physical Chemistry B* **2003**, *107* (26), 6343-6349.

134. Premadasa, U. I.; Adhikari, N. M.; Cimatu, K. L. A., Molecular Insights into the Role of Electronic Substituents on the Chemical Environment of the −CH3 and >C=O Groups of Neat Liquid Monomers Using Sum Frequency Generation Spectroscopy. *The Journal of Physical Chemistry C* **2019**, *123* (46), 28201-28209.

135. Malyk, S.; Shalhout, F. Y.; O'Leary, L. E.; Lewis, N. S.; Benderskii, A. V., Vibrational Sum Frequency Spectroscopic Investigation of the Azimuthal Anisotropy and Rotational Dynamics of Methyl-Terminated Silicon(111) Surfaces. *J Phys Chem C* **2013**, *117* (2), 935-944.

136. Vericat, C.; Vela, M. E.; Benitez, G.; Carro, P.; Salvarezza, R. C., Self-assembled monolayers of thiols and dithiols on gold: new challenges for a well-known system. *Chemical Society Reviews* **2010**, *39* (5), 1805-1834.

137. Miranda, P. B.; Pflumio, V.; Saijo, H.; Shen, Y. R., Surfactant monolayers at solid–liquid interfaces: conformation and interaction. *Thin Solid Films* **1998**, *327-329*, 161-165.

138. Premadasa, U. I.; Adhikari, N. M.; Baral, S.; Aboelenen, A. M.; Cimatu, K. L. A., Conformational Changes of Methacrylate-Based Monomers at the Air-Liquid Interface Due to Bulky Substituents. *J Phys Chem C* **2017**, *121* (31), 16888-16902.

139. Chan, S. C.; Jang, J. H.; Cimatu, K. A., Orientational Analysis of Interfacial Molecular Groups of a 2-Methoxyethyl Methacrylate Monomer Using Femtosecond Sum Frequency Generation Spectroscopy. *J Phys Chem C* **2016**, *120* (51), 29358-29373.

140. Meena, S. K.; Sulpizi, M., Understanding the Microscopic Origin of Gold Nanoparticle Anisotropic Growth from Molecular Dynamics Simulations. *Langmuir* **2013**, *29* (48), 14954-14961.

141. da Silva, J. A.; Meneghetti, M. R., New Aspects of the Gold Nanorod Formation Mechanism via Seed-Mediated Methods Revealed by Molecular Dynamics Simulations. *Langmuir* **2018**, *34* (1), 366-375.

142. Patel, A. J.; Varilly, P.; Chandler, D., Fluctuations of Water near Extended Hydrophobic and Hydrophilic Surfaces. *J. Phys. Chem. B* **2010**, *114* (4), 1632-1637.

143.   Liu, X.; Chen, S.; Tian, F.; Ma, H.; Shen, L.; Zhai, H., Studies of protection of iron corrosion by rosin imidazoline self-assembled monolayers. *Surface and Interface Analysis: An International Journal devoted to the development and application of techniques for the analysis of surfaces, interfaces and thin films* **2007,** *39* (4), 317-323.

144.   Bhushan, B., *Nanotribology and nanomechanics: an introduction.* Springer Science & Business Media: **2008.**

145.   Vericat, C.; Vela, M. E.; Benitez, G. A.; Gago, J. A. M.; Torrelles, X.; Salvarezza, R. C., Surface characterization of sulfur and alkanethiol self-assembled monolayers on Au (111). *Journal of Physics: Condensed Matter* **2006,** *18* (48), R867.

146.   Daniel, M.-C.; Astruc, D., Gold nanoparticles: assembly, supramolecular chemistry, quantum-size-related properties, and applications toward biology, catalysis, and nanotechnology. *Chemical reviews* **2004,** *104* (1), 293-346.

147.   de Alwis, C.; Leftwich, T. R.; Perrine, K. A., New Approach to Simultaneous In Situ Measurements of the Air/Liquid/Solid Interface Using PM-IRRAS. *Langmuir* **2020,** *36* (13), 3404-3414.

148.   Lewis, P. A.; Donhauser, Z. J.; Mantooth, B. A.; Smith, R. K.; Bumm, L. A.; Kelly, K. F.; Weiss, P. S., Control and placement of molecules via self-assembly. *Nanotechnology* **2001,** *12* (3), 231.

149.   Cimatu, K.; Baldelli, S., Chemical imaging of corrosion: Sum frequency generation imaging microscopy of cyanide on gold at the solid-liquid interface. *J Am Chem Soc* **2008,** *130* (25), 8030-8037.

150.   Spencer, W.; Savini, B.; Somnath, B.; L. Robert, B., *Momentum Relaxed, Plasmon-Resonant Vibrational Sum Frequency Generation of Electrochemical Interfaces: Direct Observation of Carbon Dioxide Electroreduction on Gold.* **2020.**

151.   Patel, A. J.; Varilly, P.; Chandler, D., Fluctuations of water near extended hydrophobic and hydrophilic surfaces. *The Journal of Physical Chemistry B* **2010,** *114* (4), 1632-1637.

152.   Frisch, M. J.; Trucks, G. W.; Schlegel, H. B.; Scuseria, G. E.; Robb, M. A.; Cheeseman, J. R.; Scalmani, G.; Barone, V.; Petersson, G. A.; Nakatsuji, H.; Li, X.; Caricato, M.; Marenich, A. V.; Bloino, J.; Janesko, B. G.; Gomperts, R.; Mennucci, B.; Hratchian, H. P.; Ortiz, J. V.; Izmaylov, A. F.; Sonnenberg, J. L.; Williams; Ding, F.; Lipparini, F.; Egidi, F.; Goings, J.; Peng, B.; Petrone, A.; Henderson, T.; Ranasinghe, D.; Zakrzewski, V. G.; Gao, J.; Rega, N.; Zheng, G.; Liang, W.; Hada, M.; Ehara, M.; Toyota, K.; Fukuda, R.; Hasegawa, J.; Ishida, M.; Nakajima, T.; Honda, Y.; Kitao, O.; Nakai, H.; Vreven, T.; Throssell, K.; Montgomery Jr., J. A.; Peralta, J. E.; Ogliaro, F.; Bearpark, M. J.; Heyd, J. J.; Brothers, E. N.; Kudin, K. N.; Staroverov, V. N.; Keith, T. A.; Kobayashi, R.; Normand, J.; Raghavachari, K.; Rendell, A. P.; Burant, J. C.; Iyengar, S. S.; Tomasi, J.; Cossi, M.; Millam, J. M.; Klene, M.;

Adamo, C.; Cammi, R.; Ochterski, J. W.; Martin, R. L.; Morokuma, K.; Farkas, O.; Foresman, J. B.; Fox, D. J. *Gaussian 16 Rev. C.01*, Wallingford, CT, 2016.

153. Wang, J.; Wolf, R. M.; Caldwell, J. W.; Kollman, P. A.; Case, D. A., Development and testing of a general amber force field. *Journal of computational chemistry* **2004,** *25* (9), 1157-1174.

154. Joung, I. S.; Cheatham III, T. E., Determination of alkali and halide monovalent ion parameters for use in explicitly solvated biomolecular simulations. *The journal of physical chemistry B* **2008,** *112* (30), 9020-9041.

155. Berendsen, H.; Grigera, J.; Straatsma, T., The missing term in effective pair potentials. *Journal of Physical Chemistry* **1987,** *91* (24), 6269-6271.

156. Heinz, H.; Vaia, R.; Farmer, B.; Naik, R., Accurate simulation of surfaces and interfaces of face-centered cubic metals using 12− 6 and 9− 6 Lennard-Jones potentials. *The Journal of Physical Chemistry C* **2008,** *112* (44), 17281-17290.

157. Plimpton, S., Fast parallel algorithms for short-range molecular dynamics. *Journal of computational physics* **1995,** *117* (1), 1-19.

APPENDIX A: SUPPORTING DETAILS FOR CHAPTER 3

^{1}H NMR in D$_2$O: δ = 7.61 (Ph, dt, J = 8.7, 4.2 Hz, 1H), 7.56 (Ph, d, J = 4.4 Hz, 4H), 4.49 (Ph-CH$_2$, s, 2H), 3.32 – 3.25 (m, 2H), 3.03 (s, 6H), 1.86 (td, J = 12.1, 10.2, 6.1 Hz, 2H), 1.40 (h, J = 7.4 Hz, 2H), 0.98 (t, J = 7.4 Hz, 3H).

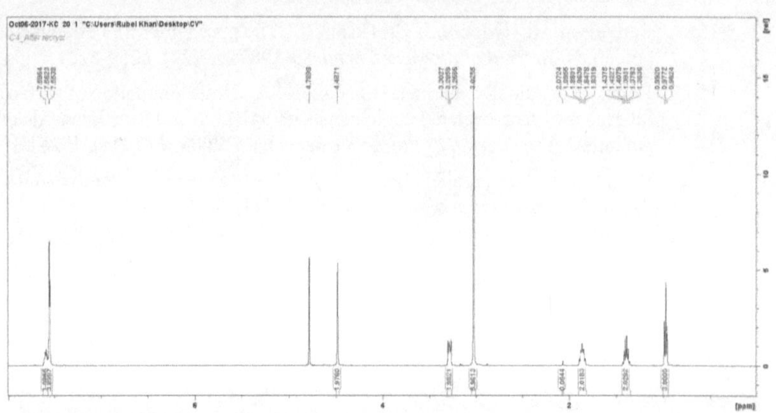

Figure A.1. ^{1}H-NMR spectrum of benzyldimethyloctylammonium bromide (C4)

^{1}H NMR in CDCl$_3$: δ = 7.66 (Ph, d, 2H), 7.44 (Ph, d, 3H, J = 7.1 Hz), 5.09 (Ph-CH$_2$, s, 2H), 3.54 (NCH$_2$, t, 2H, J = 8.2 Hz), 3.31 (N-CH$_3$, s, 6H), 1.79 (CH$_2$, d, 2H), 1.30 (CH$_2$, t, 6H, J = 17.2 Hz), 0.86 (C-CH$_3$), t, 3H, J = 6.9 Hz

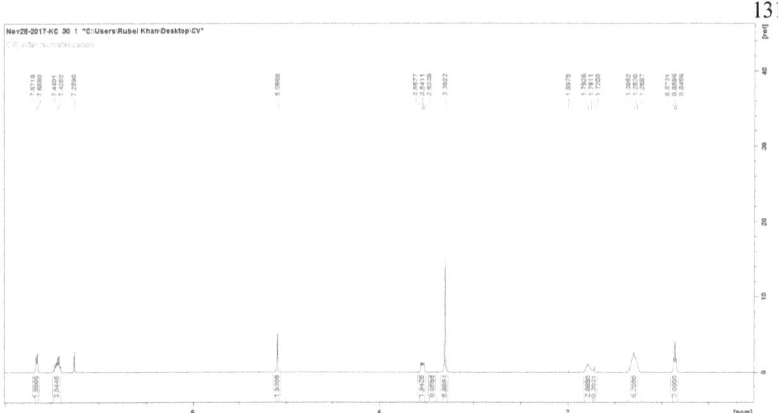

Figure A.2. ^{1}H-NMR spectrum of benzyldimethyloctylammonium bromide (C6)

^{1}H NMR in CDCl$_3$: δ = 7.62 (Ph, d, 2H), 7.37 (Ph, m, 3H, J = 9.0 Hz), 5.03 (Ph-CH$_2$, t, 2H), 3.51 (NCH$_2$, t, 2H, J = 8.2 Hz), 3.10 (N-CH$_3$, s, 6H), 1.96 (CH$_2$, s, 2H), 1.21 (CH$_2$, d, 14H, J = 13.7 Hz), 0.84 (C-CH$_3$), t, 3H, J = 6.7 Hz

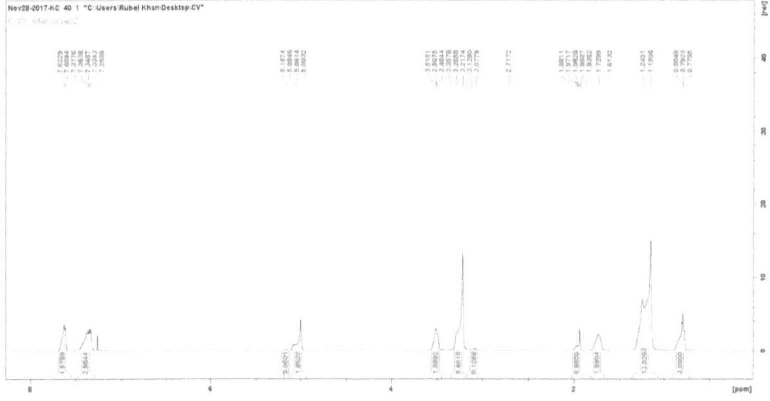

Figure A.3. ^{1}H-NMR spectrum of benzyldimethyloctylammonium bromide (C10)

^{1}H NMR in CDCl$_3$: δ = 7.61 (Ph, d, 2H), 7.35 (Ph, m, 3H, J = 7 Hz), 4.99 (Ph-CH$_2$, s, 2H), 3.49 (NCH$_2$, t, 2H, J = 8.2 Hz), 3.21 (N-CH$_3$, s, 6H), 1.71 (CH$_2$, s, 2H), 1.19 (CH$_2$, m, 18H, J = 14.4 Hz), 0.79 (C-CH$_3$), t, 3H, J = 6.9 Hz

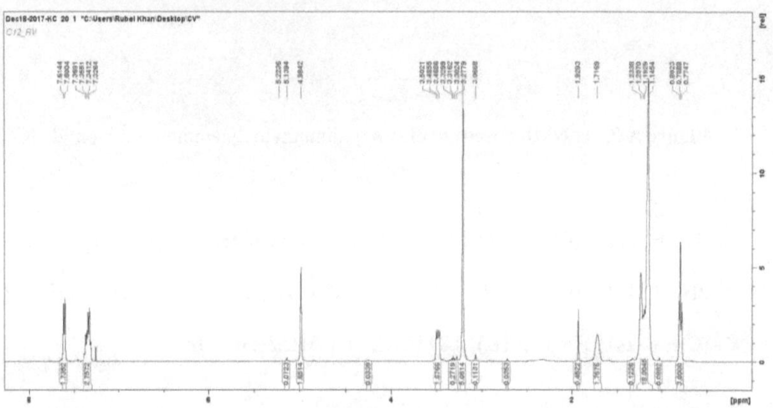

Figure A.4. ^{1}H-NMR spectrum of benzyldimethyloctylammonium bromide (C12)

APPENDIX B: SUPPORTING DETAILS FOR CHAPTER 4

Role of the cationic headgroup to conformational changes undergone by shorter alkyl chain surfactant and water molecules at the air-liquid interface

Table B.1. The purity of Quats determined by ^{1}H-NMR.

Quats	Purity (%)
C4	99.71
C6	99.63
C8	99.67
C10	99.27
C12	99.028

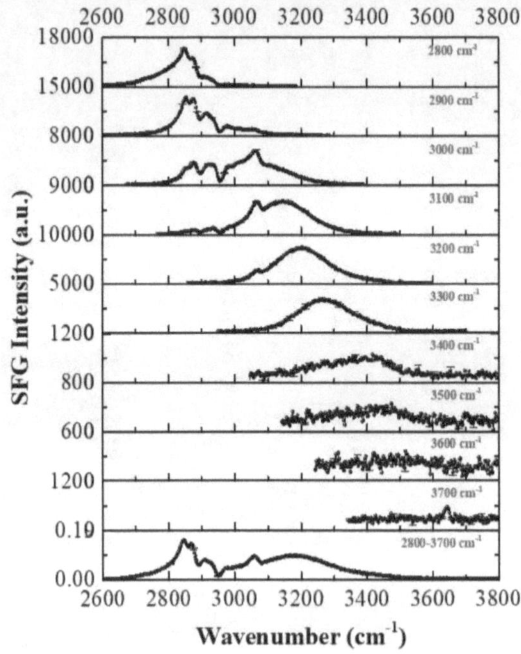

Figure: B.1. This is a representation of the data collection and processing of one Quat compound (C8) at ssp polarization. The spectra were collected from 2800 cm^{-1} to 3700 cm^{-1}, summed up, corrected for the polarization efficiency of the optics, and normalized with the reference sample eicosanoic acid.

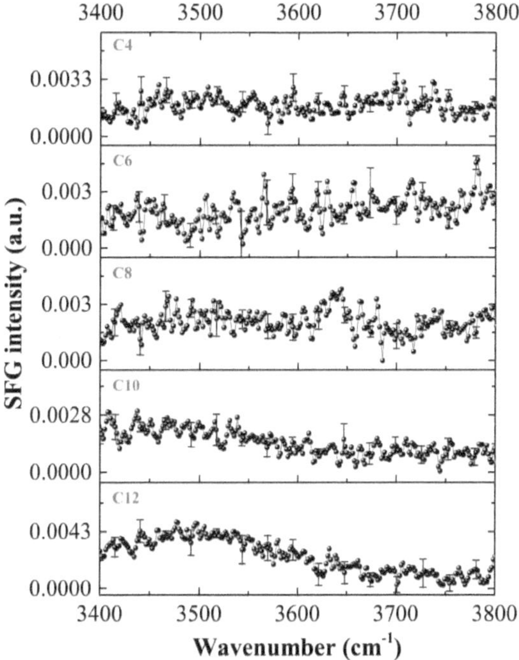

Figure B.2. This is a representation of the free OH region at ~ 3700 cm^{-1} of five Quat compounds at ssp polarization. The spectra from 3600 cm^{-1} and 3700 cm^{-1} are summed up, corrected for the polarization efficiency of the optics, and normalized with the reference sample eicosanoic acid.

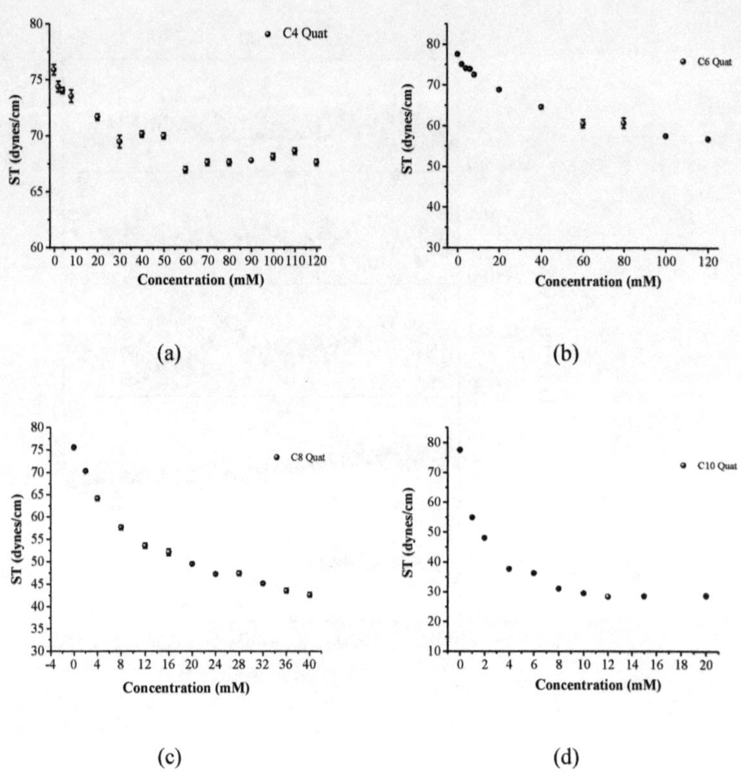

(a)

(b)

(c)

(d)

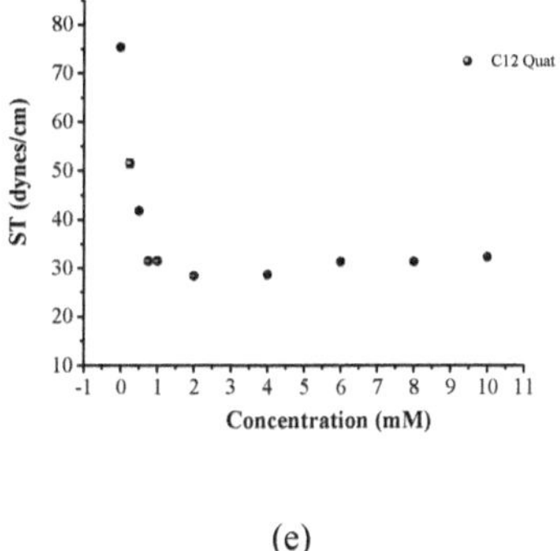

(e)

Figure B.3. Surface tension of (a) C4, (b) C6, (c) C8, (d) C10, and (e) C12 Quat compounds.

Table B.2. Surface excess concentration for Quat compounds.

C mM	Γ_{C4} molecules/cm^2	C mM	Γ_{C6} molecules/cm^2	C mM	Γ_{C8} molecules/cm^2	C mM	Γ_{C10} molecules/cm^2	C mM	Γ_{C12} molecules/cm^2
0	--	0	--	0	--	0	--	0	--
2	1.88×10^{13}	2	3.12×10^{13}	2	6.43×10^{13}	1	2.79×10^{14}	0.25	2.94×10^{14}
4	2.30×10^{13}	4	4.37×10^{13}	4	1.41×10^{14}	2	3.63×10^{14}	0.5	4.13×10^{14}
8	2.92×10^{13}	6	4.58×10^{13}	8	2.20×10^{14}	4	4.91×10^{14}	0.75	5.39×10^{14}
20	5.19×10^{13}	8	6.22×10^{13}	12	2.69×10^{14}	6	5.07×10^{14}	1	5.38×10^{14}
30	7.89×10^{13}	20	1.08×10^{14}	16	2.88×10^{14}	8	5.71×10^{14}	2	5.76×10^{14}
40	7.07×10^{13}	40	1.60×10^{14}	20	3.20×10^{14}	10	5.91×10^{14}	4	5.73×10^{14}
50	7.28×10^{13}	60	2.10×10^{14}	24	3.48×10^{14}	12	6.04×10^{14}	6	5.41×10^{14}
60	1.10×10^{14}	80	2.08×10^{14}	28	3.46×10^{14}	15	6.03×10^{14}	8	5.41×10^{14}
70	1.02×10^{14}	100	2.47×10^{14}	32	3.74×10^{14}	20	6.00×10^{14}	10	5.29×10^{14}
80	1.02×10^{14}	120	2.55×10^{14}	36	3.93×10^{14}				
90	9.96×10^{13}			40	4.05×10^{14}				
100	9.55×10^{13}								
110	8.93×10^{13}								
120	1.02×10^{14}								

Table B.3.1. Fitting parameters for 8 mM C4 Quat with 0% NaCl in H_2O at ssp polarization.

Parameters	Estimated Value	Standard Error	Error with 95% Confidence Level
A1	0	0	0
A2	2.5	0.2	0.3
A3	3.7	0.2	0.4
A4	2.1	0.2	0.4
A5	2.5	0.2	0.4
A6	1.3	0.3	0.6
A7	1.2	0.2	0.3
A8	21.7	0.7	1.3
A9	4.3	0.9	1.7
$\Gamma1$	13.5	7.8×10^{-14}	1.5×10^{-13}
$\Gamma2$	18	0.6	1.2
$\Gamma3$	17.3	0.6	1.2
$\Gamma4$	16	1.0	2.0
$\Gamma5$	15.6	0.8	1.5
$\Gamma6$	30	4.9	9.8
$\Gamma7$	20	1.9	3.83
$\Gamma8$	104.5	1.7	3.45
$\Gamma9$	97.4	13.5	27.0
$\omega1$	2718.4	7.4×10^{-15}	1.5×10^{-14}
$\omega2$	2850	0.3	0.6
$\omega3$	2877.6	0.2	0.4
$\omega4$	2910	0.3	0.7
$\omega5$	2936.2	0.3	0.6
$\omega6$	2986	2.1	4.1
$\omega7$	3062	0.9	1.9
$\omega8$	3178.4	2.7	5.4
$\omega9$	3443.3	6.6	13.1
n	0	0	0
p	2	0.1	0.2
n1	0	0	0
R^2	0.99		

Table B.3.2. Fitting parameters for 8 mM C4 Quat with 1% NaCl in H_2O at ssp polarization.

Parameters	Estimated Value	Standard Error	Error with 95% Confidence Level
A1	0.1	0.1	0.2
A2	0.6	0.1	0.2
A3	2.8	0.3	0.5
A4	3.1	0.9	1.8
A5	1.6	1.6	3.2
A6	0.1	0.9	1.9
A7	0.8	0.3	0.7
A8	14.3	4.8	9.6
A9	5	2.4	4.9
$\Gamma 1$	24.9	18.8	37.6
$\Gamma 2$	10.3	0.9	1.9
$\Gamma 3$	17.3	1.1	2.1
$\Gamma 4$	22.9	3.5	6.9
$\Gamma 5$	15.9	6.9	13.8
$\Gamma 6$	11.9	3538.6	7077.3
$\Gamma 7$	25	7.7	15.3
$\Gamma 8$	150	29.3	58.6
$\Gamma 9$	100	19.6	39.2
$\omega 1$	2730	9.2	18.4
$\omega 2$	2854.7	0.4	0.9
$\omega 3$	2879	0.4	0.7
$\omega 4$	2920.2	1.9	3.9
$\omega 5$	2947.4	3.7	7.5
$\omega 6$	2961.9	2400.6	4801.3
$\omega 7$	3050.4	2.8	5.6
$\omega 8$	3150	8.7	17.3
$\omega 9$	3310.1	7.9	15.8
n	0	0	0
p	1	0.1	0.2
n1	1.5×10^{-15}	0	0

R²	0.99		

Table B.3.3. Fitting parameters for 8 mM C4 Quat with 10% NaCl in H_2O at ssp polarization.

Parameters	Estimated Value	Standard Error	Error with 95% Confidence Level
A1	0	0	0
A2	1	0.1	0.2
A3	3.1	0.2	0.4
A4	2.3	0.3	0.5
A5	1.7	0.2	0.3
A6	0	0	0
A7	0.9	0.2	0.3
A8	12.2	1.7	3.4
A9	3.1	0.9	1.9
$\Gamma 1$	11.4	47.3	94.6
$\Gamma 2$	12.5	0.9	1.9
$\Gamma 3$	16.2	0.6	1.2
$\Gamma 4$	19.5	1.5	3
$\Gamma 5$	13.8	0.8	1.6
$\Gamma 6$	3.1	2.4	4.9
$\Gamma 7$	21.4	3.4	6.7
$\Gamma 8$	141.8	12.4	24.8
$\Gamma 9$	100	17.7	35.4
$\omega 1$	2730	27.2	54.5
$\omega 2$	2853.9	0.4	0.8
$\omega 3$	2878.9	0.2	0.4
$\omega 4$	2917.5	0.6	1.2
$\omega 5$	2944.6	0.3	0.7
$\omega 6$	3010	1.6	3.2
$\omega 7$	3076.5	1.7	3.5
$\omega 8$	3232.5	4.2	8.4
$\omega 9$	3442.8	6.9	14
n	0	0	0

p	-0.1	1.4	2.7
n1	0	0	0
R²	0.99		

Table B.3.4. Fitting parameters for 8 mM C6 Quat with 0% NaCl in H_2O at ssp polarization.

Parameters	Estimated Value	Standard Error	Error with 95% Confidence Level
A1	0	0.1	0.1
A2	1.2	0.2	0.4
A3	2.3	0.3	0.6
A4	1.6	0.3	0.7
A5	0.9	0.2	0.4
A6	4.3×10^{-05}	0.3	0.6
A7	0.8	0.2	0.4
A8	9.8	0.7	1.4
A9	0.3	0.6	1.2
$\Gamma1$	15.1	220087.4	440174.8
$\Gamma2$	12	1.3	2.6
$\Gamma3$	14.5	1.2	2.3
$\Gamma4$	18.2	2.7	5.4
$\Gamma5$	10.1	1.5	3
$\Gamma6$	14.8	91922.5	183844.9
$\Gamma7$	14.8	3.2	6.4
$\Gamma8$	93.1	6.3	12.5
$\Gamma9$	50	86.6	173.2
$\omega1$	2716.7	132277.5	264555
$\omega2$	2855.2	0.7	1.3
$\omega3$	2877.9	0.5	0.9
$\omega4$	2919.3	1.2	2.4
$\omega5$	2941.8	0.6	1.2
$\omega6$	2975.9	42523	85046
$\omega7$	3045.2	1.9	3.8
$\omega8$	3170	7.3	14.6

ω9	3450	37	74
n	0.01	0.01	0.01
p	1.7	0.1	0.3
n1	0	0	0
R^2	0.97		

Table B.3.5. Fitting parameters for 8 mM C6 Quat with 1% NaCl in H_2O at ssp polarization.

Parameters	Estimated Value	Standard Error	Error with 95% Confidence Level
A1	0	0	0
A2	0.9	0.1	0.2
A3	3.1	0.2	0.3
A4	2.3	0.2	0.5
A5	1.7	0.1	0.3
A6	0	0.1	0.1
A7	0.9	0.2	0.3
A8	12.2	1.7	3.4
A9	3.1	0.9	1.8
Γ1	11.4	47.2	94.5
Γ2	12.5	0.9	1.8
Γ3	16.2	0.6	1.2
Γ4	19.5	1.5	2.9
Γ5	13.8	0.8	1.6
Γ6	3.1	2.4	4.8
Γ7	21.4	3.3	6.7
Γ8	141.8	12.4	24.7
Γ9	100	17.7	35.4
ω1	2730	27.2	54.4
ω2	2854	0.4	0.8
ω3	2878.9	0.2	0.4
ω4	2917.5	0.6	1.2
ω5	2944.6	0.3	0.7
ω6	3010	1.6	3.2

ω7	3076.5	1.7	3.5
ω8	3232.5	4.2	8.3
ω9	3442.8	7	13.9
n	0	0	0
p	-0.0	1.3	2.7
n1	0	0	0
R^2	0.99		

Table B.3.6. Fitting parameters for 8 mM C6 Quat with 10% NaCl in H_2O at ssp polarization

Parameters	Estimated Value	Standard Error	Error with 95% Confidence Level
A1	0	0	0
A2	1.3	0.2	0.5
A3	2.4	0.3	0.6
A4	1	0.4	0.8
A5	2.1	0.5	1
A6	1.8	0.4	0.9
A7	2.3	0.1	0.3
A8	14.7	4.4	8.8
A9	9.5	3.7	7.5
$\Gamma1$	14.1	7.6×10^{-14}	1.5×10^{-13}
$\Gamma2$	15	1.3	2.7
$\Gamma3$	15	1.1	2.1
$\Gamma4$	15	3.4	6.8
$\Gamma5$	19	2.9	5.8
$\Gamma6$	25	4.4	8.8
$\Gamma7$	16.2	0.9	1.7
$\Gamma8$	126.7	17.3	34.6
$\Gamma9$	121.9	21.9	43.7
ω1	2717.4	2.9×10^{-14}	5.9×10^{-05}
ω2	2853.1	0.6	1.3
ω3	2875.4	0.3	0.7
ω4	2910	1.6	3.2
ω5	2935.5	1.3	2.6

$\omega 6$	2979.9	1.8	3.7
$\omega 7$	3062.9	0.6	1.1
$\omega 8$	3200	7	14
$\omega 9$	3359.5	9.5	19.
n	0.03	0.01	0.02
p	1.7	0.1	0.2
n1	0	0	0
R^2	0.98		

Table B.3.7. Fitting parameters for 8 mM C8 Quat with 0% NaCl in H_2O at ssp

polarization

Parameters	Estimated Value	Standard Error	Error with 95% Confidence Level
A1	0	0	0
A2	3.8	0.1	0.3
A3	3.5	0.2	0.3
A4	2.9	0.2	0.3
A5	0.7	0.1	0.2
A6	3.8	0.2	0.4
A7	6.8	0.3	0.6
A8	22.9	0.4	0.8
A9	1.6×10^{-08}	0.2	0.4
$\Gamma 1$	14	368099.8	736199.6
$\Gamma 2$	18	0.4	0.8
$\Gamma 3$	15	0.4	0.8
$\Gamma 4$	17.7	0.7	1.4
$\Gamma 5$	8.6	0.5	1.1
$\Gamma 6$	35	1.3	2.5
$\Gamma 7$	40	1	1.9
$\Gamma 8$	88	1.4	2.7
$\Gamma 9$	96.3	7.0×10^7	1.4×10^8
$\omega 1$	2719.6	60547	121094.
$\omega 2$	2845.1	0.2	0.4
$\omega 3$	2871.7	0.1	0.3

$\omega 4$	2910.2	0.3	0.6
$\omega 5$	2929.9	0.2	0.4
$\omega 6$	2981	0.5	1
$\omega 7$	3055	0.3	0.7
$\omega 8$	3165	0.9	1.8
$\omega 9$	3449.8	4.5×10^{8}	9.0×10^{8}
n	0.02	0	0
p	0.2	0.2	0.4
n1	0	0	0
R^2	0.98		

Table B.3.8. Fitting parameters for 8 mM C8 Quat with 0.5% NaCl in H_2O at ssp

polarization

Parameters	Estimated Value	Standard Error	Error with 95% Confidence Level
A1	0	0	0
A2	2.7	0.2	0.4
A3	2.6	0.2	0.4
A4	1.6	0.2	0.3
A5	0.7	0.1	0.2
A6	0.3	0.1	0.1
A7	1.3	0.2	0.3
A8	18.9	1.1	2.1
A9	3.5	0.4	0.8
$\Gamma 1$	13.2	8.2×10^{-15}	1.6×10^{-14}
$\Gamma 2$	18	0.8	1.6
$\Gamma 3$	14.6	0.7	1.4
$\Gamma 4$	16.5	1.3	2.6
$\Gamma 5$	9.4	0.8	1.7
$\Gamma 6$	11.4	3.1	6.1
$\Gamma 7$	30	2.9	5.9
$\Gamma 8$	120.3	3.9	7.9
$\Gamma 9$	68.1	5	10
$\omega 1$	2718.2	2.9×10^{-15}	5.8×10^{-15}

$\omega 2$	2849.9	0.4	0.8
$\omega 3$	2876.2	0.2	0.5
$\omega 4$	2914.9	0.6	1.2
$\omega 5$	2935.3	0.4	0.8
$\omega 6$	2980	1.4	2.7
$\omega 7$	3061.4	1.2	2.5
$\omega 8$	3206	1.5	3.1
$\omega 9$	3380	2.4	4.8
n	0	0	0
p	2.5	0.6	1.1
n1	0	0	0
R^2	0.98		

Table B.3.9. Fitting parameters for 8 mM C8 Quat with 1% NaCl in H_2O at ssp polarization

Parameters	Estimated Value	Standard Error	Error with 95% Confidence Level
A1	0	0	0
A2	3.6	0.2	0
A3	1.9	0.1	0.2
A4	1.7	0.1	0.2
A5	0.7	0.06	0.12
A6	0.2	0.07	0.13
A7	1.3	0.05	0.10
A8	19.1	0.5	0.9
A9	0.5	0.06	0.1
$\Gamma 1$	10	21.2	42.3
$\Gamma 2$	22.5	0.6	1.3
$\Gamma 3$	11.3	0.3	0.6
$\Gamma 4$	17.8	0.8	1.5
$\Gamma 5$	9.5	0.4	0.9
$\Gamma 6$	14.8	3.3	6.6
$\Gamma 7$	17.6	0.5	1.1
$\Gamma 8$	129.2	1.5	3
$\Gamma 9$	26.6	3	6

	Estimated Value	Standard Error	Error with 95% Confidence Level
$\omega 1$	2730	12	24.6
$\omega 2$	2853.8	0.3	0.7
$\omega 3$	2877	0.1	0.2
$\omega 4$	2915	0.4	0.8
$\omega 5$	2935	0.2	0.4
$\omega 6$	2986	1.5	3
$\omega 7$	3057.6	0.2	0.5
$\omega 8$	3190.2	1.4	2.7
$\omega 9$	3400	1.9	3.7
n	0	0	0
p	2.5×10^{-06}	5.6	11.3
n1	0	5.3×10^{-5}	0
R^2	0.99		

Table B.3.10. Fitting parameters for 8 mM C8 Quat with 5% NaCl in H_2O at ssp polarization

Parameters	Estimated Value	Standard Error	Error with 95% Confidence Level
A1	0	0	0
A2	2	0.1	0.2
A3	3.3	0.1	0.3
A4	3.3	0.4	0.8
A5	1	0.2	0.4
A6	0.4	0.2	0.4
A7	1	0.1	0.2
A8	10	1.5	3.1
A9	7.5	1.5	3.1
$\Gamma 1$	8.4	2.1×10^{-14}	4.2×10^{-14}
$\Gamma 2$	15	0.5	1
$\Gamma 3$	13.6	0.3	0.7
$\Gamma 4$	21.9	1.8	3.7
$\Gamma 5$	9.2	1	2
$\Gamma 6$	12.5	5.7	11.4
$\Gamma 7$	10	0.9	1.8

$\Gamma 8$	96.7	10.2	20.4
$\Gamma 9$	100	12.7	25.4
$\omega 1$	2719.7	4.1×10^{-15}	8.1×10^{-15}
$\omega 2$	2849.6	0.2	0.5
$\omega 3$	2874.3	0.1	0.3
$\omega 4$	2914.8	1.1	2.3
$\omega 5$	2934.5	0.3	0.7
$\omega 6$	2980	3.2	6.5
$\omega 7$	3058.1	0.6	1.3
$\omega 8$	3150	4.3	8.6
$\omega 9$	3341.1	6.2	12.3
n	0.04	0	0
p	1.9	0.04	0.08
n1	0	0	0
R^2	0.99		

Table B.3.11. Fitting parameters for 8 mM C8 Quat with 10% NaCl in H_2O at ssp polarization

Parameters	Estimated Value	Standard Error	Error with 95% Confidence Level
A1	0	0	0
A2	2.3	0.02	0.05
A3	3.1	0.04	0.08
A4	3.6	0.12	0.24
A5	0.4	0.05	0.11
A6	0.1	0.01	0.02
A7	0.8	0.02	0.03
A8	9	0.7	1.34
A9	3.3	0.6	1.1
$\Gamma 1$	7.6	23.7	47.3
$\Gamma 2$	15	0.1	0.2
$\Gamma 3$	14.9	0.1	0.2
$\Gamma 4$	29.7	0.5	1
$\Gamma 5$	9.2	0.7	1.5
$\Gamma 6$	4.2	0.7	1.4

	Estimated Value	Standard Error	Error with 95% Confidence Level
$\Gamma 7$	6.9	0.2	0.3
$\Gamma 8$	95.5	3.8	7.7
$\Gamma 9$	70.3	6.7	13.5
$\omega 1$	2720.7	15.3	30.6
$\omega 2$	2849.7	0.1	0.1
$\omega 3$	2875.4	0.04	0.08
$\omega 4$	2919.5	0.5	1.0
$\omega 5$	2936.9	0.2	0.4
$\omega 6$	2975.6	0.4	0.8
$\omega 7$	3058.7	0.1	0.2
$\omega 8$	3222.6	1.9	3.8
$\omega 9$	3351.7	2.5	5.1
n	0.01	0	0
p	1.1	0.04	0.08
$n1$	2.0×10^{-06}	2.0×10^{-05}	4.0×10^{-05}
R^2	0.99		

Table B.3.12. Fitting parameters for 8 mM C10 Quat with 0% NaCl in H_2O at ssp

polarization

Parameters	Estimated Value	Standard Error	Error with 95% Confidence Level
A1	0	0	0
A2	3.5	0.2	0.3
A3	4.5	0.2	0.3
A4	2.6	0.2	0.3
A5	1.3	0.1	0.2
A6	1.3	0.1	0.2
A7	3.9	0.3	0.6
A8	29.4	0.4	0.8
A9	0.08	0.03	0.06
$\Gamma 1$	30.4	0	0
$\Gamma 2$	20	0.5	1.1
$\Gamma 3$	14.2	0.3	0.6
$\Gamma 4$	15.8	0.8	1.5
$\Gamma 5$	8.2	0.4	0.8

	Estimated Value	Standard Error	Error with 95% Confidence Level
$\Gamma 6$	15	0.9	1.9
$\Gamma 7$	28	1.5	2.9
$\Gamma 8$	98	1.4	2.8
$\Gamma 9$	12	4.7	9.5
$\omega 1$	2719.3	0	0
$\omega 2$	2848.5	0.3	0.6
$\omega 3$	2876.2	0.1	0.2
$\omega 4$	2915.8	0.4	0.9
$\omega 5$	2935.6	0.2	0.4
$\omega 6$	2980	0.4	0.8
$\omega 7$	3055	0.6	1.3
$\omega 8$	3166.5	1	2
$\omega 9$	3396	2.8	5.6
n	0.01	0	0
p	3.7×10^{-05}	0.3	0.6
$n1$	0	0	0
R^2	0.97		

Table B.3.13. Fitting parameters for 8 mM C10 Quat with 1% NaCl in H_2O at ssp polarization

Parameters	Estimated Value	Standard Error	Error with 95% Confidence Level
A1	0	0	0
A2	2.7	0.1	0.2
A3	3.7	0.1	0.2
A4	1.8	0.1	0.2
A5	0.7	0.5	0.1
A6	0.3	0	0.1
A7	1.4	0.1	0.1
A8	22.1	0.3	0.7
A9	0.9	0.1	0.1
$\Gamma 1$	26.5	0	0
$\Gamma 2$	18	0.3	0.7
$\Gamma 3$	14.8	0.2	0.5
$\Gamma 4$	15	0.5	1.1

$\Gamma 5$	7.4	0.3	0.6
$\Gamma 6$	10	1	2
$\Gamma 7$	18	0.6	1.2
$\Gamma 8$	128.4	1.4	2.8
$\Gamma 9$	30.1	2.3	4.5
$\omega 1$	2730.2	0	0
$\omega 2$	2846.5	0.2	0.3
$\omega 3$	2874.4	0.1	0.2
$\omega 4$	2915.4	0.3	0.5
$\omega 5$	2935	0.1	0.3
$\omega 6$	2978	0.5	1
$\omega 7$	3055.7	0.3	0.6
$\omega 8$	3188.4	1	1.9
$\omega 9$	3390	1.2	2.4
n	0	0	0
p	1	0.1	0.2
n1	0	0	0
R^2	0.97		

Table B.3.14. Fitting parameters for 8 mM C10 Quat with 10% NaCl in H_2O at ssp

polarization

Parameters	Estimated Value	Standard Error	Error with 95% Confidence Level
A1	0	0	0
A2	1.7	0.1	0.3
A3	2.1	0.3	0.5
A4	3.1	0.6	1.1
A5	0.3	0.2	0.4
A6	1.5	0.3	0.6
A7	2.1	0.1	0.2
A8	9.2	2.2	4.3
A9	11	3.4	6.9
$\Gamma 1$	15.3	122259	244518
$\Gamma 2$	15	0.9	1.7

	Estimated Value	Standard Error	Error with 95% Confidence Level
$\Gamma 3$	15	1.1	2.2
$\Gamma 4$	25	3.0	6
$\Gamma 5$	8.8	2.9	5.7
$\Gamma 6$	25	3.3	6.5
$\Gamma 7$	17.4	0.9	1.8
$\Gamma 8$	101.2	11.1	22.1
$\Gamma 9$	153.4	30.2	60.3
$\omega 1$	2732.8	64069.8	128139.5
$\omega 2$	2851.1	0.4	0.9
$\omega 3$	2875.2	0.3	0.7
$\omega 4$	2916.5	1.6	3.2
$\omega 5$	2936.7	0.9	1.7
$\omega 6$	2975.7	1.3	2.6
$\omega 7$	3057.8	0.6	1.1
$\omega 8$	3173	4.1	8.2
$\omega 9$	3364.9	12.7	25.5
n	0.04	0	0
p	2	0.1	0.2
$n1$	0	0	0
R^2	0.98		

Table B.3.15. Fitting parameters for 8 mM C12 Quat with 0% NaCl in H_2O at ssp polarization

Parameters	Estimated Value	Standard Error	Error with 95% Confidence Level
A1	0	0	0
A2	3.8	0.3	0.7
A3	3.1	0.3	0.6
A4	2.4	0.2	0.5
A5	1.1	0.2	0.3
A6	1.9	0.2	0.4
A7	3.7	0.2	0.3
A8	35.3	1	2.1
A9	2.5	0.2	0.4

Γ_1	31.2	42	84
Γ_2	17.2	1	2
Γ_3	12.9	0.7	1.5
Γ_4	15.7	1	2
Γ_5	11.2	1	2
Γ_6	21.7	1.4	2.8
Γ_7	21.7	0.7	1.4
Γ_8	110.9	1.3	2.6
Γ_9	57.8	3.3	6.5
ω_1	2751.2	19.5	39
ω_2	2846.7	0.4	0.8
ω_3	2872.5	0.3	0.5
ω_4	2909.6	0.5	1.1
ω_5	2931.3	0.5	1.1
ω_6	2975.4	0.5	0.9
ω_7	3055.4	0.3	0.6
ω_8	3154.8	1	2.1
ω_9	3386.1	2	4
n	0.03	0	0
p	0.1	0.3	0.5
n1	0	0	0
R^2	0.97		

Table B.3.16. Fitting parameters for 8 mM C12 Quat with 1% NaCl in H_2O at ssp

polarization

Parameters	Estimated Value	Standard Error	Error with 95% Confidence Level
A1	0	0	0
A2	2.9	0.1	0.2
A3	1.3	0.05	0.11
A4	2.5	0.1	0.2
A5	0.3	0.06	0.12
A6	0.5	0.05	0.1
A7	1.4	0.04	0.08
A8	22.6	1.0	2.1

Parameters	Estimated Value	Standard Error	Error with 95% Confidence Level
A9	4.2	0.4	0.7
$\Gamma 1$	14.5	176735.6	353471.3
$\Gamma 2$	18	0.3	0.7
$\Gamma 3$	9.8	0.3	0.5
$\Gamma 4$	21	0.8	1.6
$\Gamma 5$	7.5	1.0	1.9
$\Gamma 6$	12.7	1.0	2.0
$\Gamma 7$	13.1	0.3	0.7
$\Gamma 8$	149.9	3.6	7.1
$\Gamma 9$	73.3	4.3	8.7
$\omega 1$	2719.1	77237.7	154475.4
$\omega 2$	2853	0.1	0.3
$\omega 3$	2877.2	0.1	0.2
$\omega 4$	2915.4	0.4	0.9
$\omega 5$	2936.2	0.3	0.7
$\omega 6$	2977.3	0.4	0.9
$\omega 7$	3058.9	0.2	0.4
$\omega 8$	3193.8	1.4	2.9
$\omega 9$	3400	2.2	4.4
n	0.02	0	0
p	2.3	0.1	0.2
n1	0	0	0
R^2	0.99		

Table B.3.17. Fitting parameters for 8 mM C12 Quat with 10% NaCl in H_2O at ssp polarization

Parameters	Estimated Value	Standard Error	Error with 95% Confidence Level
A1	0	0	0
A2	2.7	0.1	0.3
A3	1.6	0.1	0.3
A4	3.2	0.2	0.5
A5	0.8	0.2	0.4
A6	1.3	0.2	0.4
A7	1.7	0.1	0.2

A8	17.1	1.5	3
A9	2.7	0.6	1
$\Gamma 1$	12.5	0	0
$\Gamma 2$	18	0.5	1.1
$\Gamma 3$	12.2	0.6	1.2
$\Gamma 4$	19.1	1.1	2.1
$\Gamma 5$	12.1	1.8	3.7
$\Gamma 6$	25	4.3	8.6
$\Gamma 7$	17.9	1.1	2.2
$\Gamma 8$	131.3	7.9	15.9
$\Gamma 9$	49.9	6	12
$\omega 1$	2715.0	0	0
$\omega 2$	2858.7	0.3	0.6
$\omega 3$	2882.6	0.2	0.4
$\omega 4$	2918.6	0.5	1
$\omega 5$	2942	0.7	1.5
$\omega 6$	2986	1.9	3.9
$\omega 7$	3064.4	0.6	1.3
$\omega 8$	3226.4	3.8	7.7
$\omega 9$	3362.8	2.9	5.9
n	0.04	0	0
p	1.9	0.05	0.1
n1	0	0	0
R^2	0.99		

Table B.3.18. Fitting parameters for 8 mM C4 Quat with 0% NaCl in H_2O at ppp polarization

Parameters	Estimated Value	Standard Error	Error with 95% Confidence Level
A1	0.5	0.1	0.2
A2	0.01	0.01	0.02
A3	0.8	0.1	0.2
A4	3.7	0.4	0.8
A5	2.5	0.2	0.5
A6	0.2	0.1	0.2

	Estimated Value	Standard Error	Error with 95% Confidence Level
A7	0.5	0.4	0.8
A8	1.1	0.6	1.1
A9	4.7	2.5	5
A10	12.6	3.1	6.3
$\Gamma 1$	25	5.6	11.2
$\Gamma 2$	1	1.2	2.3
$\Gamma 3$	12.4	1.3	2.6
$\Gamma 4$	23.6	1.9	3.9
$\Gamma 5$	12.7	0.7	1.4
$\Gamma 6$	3.5	2.6	5.3
$\Gamma 7$	14.5	12.8	25.5
$\Gamma 8$	24.9	11.6	23.2
$\Gamma 9$	85.9	31.9	63.9
$\Gamma 10$	149.9	30.3	60.6
$\omega 1$	2740	3.5	6.9
$\omega 2$	2851.9	0.9	1.8
$\omega 3$	2886.1	0.7	1.3
$\omega 4$	2923.3	0.6	1.3
$\omega 5$	2956.4	0.4	0.8
$\omega 6$	3014.5	1.7	3.4
$\omega 7$	3045.4	6.6	13.2
$\omega 8$	3117.6	5.8	11.7
$\omega 9$	3240.3	12.1	24.2
$\omega 10$	3477.9	17.5	34.9
n	0.01	0.02	0.01
p	1.4	0.1	0.3
n1	0	0	0
R^2	0.99		

Table B.3.19. Fitting parameters for 8 mM C4 Quat with 1% NaCl in H_2O at ppp polarization

Parameters	Estimated Value	Standard Error	Error with 95% Confidence Level
A1	0.01	0.07	0.14
A2	0	0	0

A3	0.9	0.4	0.8
A4	2.4	0.9	1.9
A5	4.6	0.2	0.5
A6	0.5	0.6	1.1
A7	2.0	1.5	3
A8	2.5	1.3	2.5
A9	4.1	2.3	4.6
A10	10.2	2.5	5
$\Gamma 1$	12.2	75.8	151.6
$\Gamma 2$	9.2	1.0×10^{06}	2.0×10^{06}
$\Gamma 3$	14.4	3	6
$\Gamma 4$	29.8	7.2	14.5
$\Gamma 5$	16.8	0.5	1.0
$\Gamma 6$	14.2	10.4	20.7
$\Gamma 7$	29.1	14.6	29.2
$\Gamma 8$	40	12.6	25.2
$\Gamma 9$	80	29.2	58.4
$\Gamma 10$	114.5	19.7	39.3
$\omega 1$	2739.6	45.6	91.2
$\omega 2$	2857.2	553966.7	1107933.4
$\omega 3$	2890	1.5	3.1
$\omega 4$	2917.8	3.9	7.7
$\omega 5$	2968.2	0.5	1.0
$\omega 6$	3025	4.6	9.2
$\omega 7$	3060.7	5	9
$\omega 8$	3120	7.1	14.1
$\omega 9$	3285	10.1	20.3
$\omega 10$	3472.9	11.7	23.5
n	0.02	0	0
p	1.8	0.2	0.4
n1	0	0	0
R^2	0.99		

Table B.3.20. Fitting parameters for 8 mM C4 Quat with 10% NaCl in H_2O at ppp polarization

Parameters	Estimated Value	Standard Error	Error with 95% Confidence Level
A1	0.5	0.2	0.4
A2	0.1	0.1	0.2
A3	0.4	0.1	0.2
A4	2.1	0.3	0.6
A5	4.8	0.1	0.2
A6	0.4	0.2	0.4
A7	1.5	0.4	0.8
A8	0.7	0.4	0.8
A9	0.9	0.2	0.5
A10	13.3	1	2
$\Gamma 1$	25	12.6	25.2
$\Gamma 2$	8	3.7	7.3
$\Gamma 3$	6	1.2	2.4
$\Gamma 4$	18.7	2	4
$\Gamma 5$	14.5	0.2	0.5
$\Gamma 6$	13.3	6.8	13.5
$\Gamma 7$	20	4.5	8.9
$\Gamma 8$	20	10.1	20.3
$\Gamma 9$	20	5.9	11.8
$\Gamma 10$	150	11.9	23.7
$\omega 1$	2740	6.2	12.3
$\omega 2$	2858	2.2	4.3
$\omega 3$	2890	0.4	0.9
$\omega 4$	2910	1.1	2.1
$\omega 5$	2960.5	0.3	0.6
$\omega 6$	3025	3.8	7.6
$\omega 7$	3076.2	2.2	4.4
$\omega 8$	3120	5	10
$\omega 9$	3200	4	8
$\omega 10$	3473	12.2	24.4
n	0	0	0
p	0.4	0.5	1.1
n1	0	0	0
R^2	0.98		

Table B.3.21. Fitting parameters for 8 mM C6 Quat with 0% NaCl in H_2O at ppp

polarization

Parameters	Estimated Value	Standard Error	Error with 95% Confidence Level
A1	0.4	0.2	0.3
A2	2.8×10^{-05}	0.2	0.3
A3	0.5	0.2	0.3
A4	1.3	0.3	0.7
A5	2.3	0.1	0.3
A6	0.03	0.02	0.04
A7	0.2	0.1	0.2
A8	0.3	0.3	0.5
A9	3.5	1.4	2.8
A10	3.3	1.2	2.4
$\Gamma 1$	25	7.5	15.1
$\Gamma 2$	19.2	99454.9	198909.9
$\Gamma 3$	8.1	1.8	3.6
$\Gamma 4$	20.1	4.1	8.3
$\Gamma 5$	13.0	0.6	1.2
$\Gamma 6$	1.1	1.1	2.3
$\Gamma 7$	10.7	5.2	10.4
$\Gamma 8$	18.8	12.1	24.2
$\Gamma 9$	79.8	21.9	43.9
$\Gamma 10$	145.6	45.1	90.2
$\omega 1$	2740	3.7	7.3
$\omega 2$	2857.4	53366.5	106732.9
$\omega 3$	2890	0.8	1.6
$\omega 4$	2911.5	2.6	5.1
$\omega 5$	2960.1	0.5	1.0
$\omega 6$	3012	0.8	1.6
$\omega 7$	3072.2	2.9	5.8
$\omega 8$	3122.6	5.3	10.6
$\omega 9$	3203.9	9.5	19.1
$\omega 10$	3480	20.3	40.7
n	0.06	0.01	0.02

p	4.4	0.07	0.14
n1	0	0	0
R^2	0.97		

Table B.3.22. Fitting parameters for 8 mM C6 Quat with 1% NaCl in H_2O at ppp

polarization

Parameters	Estimated Value	Standard Error	Error with 95% Confidence Level
A1	0.2	0.1	0.1
A2	0.2	0.5	0.9
A3	0	0.7	1.3
A4	1.9	0.7	1.5
A5	4.6	0.4	0.8
A6	0.7	0.9	1.9
A7	0.3	0.4	0.8
A8	0.5	0.3	0.5
A9	0.2	1	1.9
A10	6.3	2.7	5.4
$\Gamma 1$	14.7	5.1	10.2
$\Gamma 2$	27.3	28.6	57.2
$\Gamma 3$	20.02	1099.4	2198.9
$\Gamma 4$	21.2	5	10
$\Gamma 5$	17.5	1	2
$\Gamma 6$	18.1	14.9	29.8
$\Gamma 7$	4.3	2.5	5
$\Gamma 8$	2.8	1.4	2.8
$\Gamma 9$	56.6	217.6	435.2
$\Gamma 10$	34.9	15.3	30.7
$\omega 1$	2710.6	3.2	6.4
$\omega 2$	2850.6	24.3	48.8
$\omega 3$	2887.6	387.1	774.2
$\omega 4$	2923.9	2	4
$\omega 5$	2968.2	1.4	2.9
$\omega 6$	3013.7	5.3	10.6
$\omega 7$	3069.2	2.6	5.3

ω8	3116.3	1.6	3.2
ω9	3325	148.5	297.1
ω10	3409.4	45.5	91.1
n	0.09	0.01	0.02
p	1.80	0.03	0.06
n1	0	0	0
R^2	0.98		

Table B.3.23. Fitting parameters for 8 mM C6 Quat with 10% NaCl in H_2O at ppp polarization

Parameters	Estimated Value	Standard Error	Error with 95% Confidence Level
A1	1.9×10^{-07}	0.2	0.4
A2	0.05	0.5	1
A3	0.3	0.3	0.6
A4	1	0.5	1.1
A5	2	0.5	0.9
A6	1.1	0.8	1.5
A7	1.1	0.7	1.5
A8	1.5	1.5	2.9
A9	5.5	4.8	9.6
A10	19.7	5	10
Γ1	20.9	2.1×10^{07}	41474674.9
Γ2	20	124.2	248.4
Γ3	8.9	5.8	11.6
Γ4	18	7.4	14.9
Γ5	19.1	3.1	6.2
Γ6	20.8	10.2	20.4
Γ7	15	6.1	12.3
Γ8	32.1	18.9	37.8
Γ9	80	32.5	65.0
Γ10	168	23.1	46.2
ω1	2711.4	1.2×10^{07}	2.3×10^{07}
ω2	2866.8	57.5	114.9
ω3	2895	2.8	5.6

$\omega 4$	2916	3.7	7.4
$\omega 5$	2961.2	3.1	6.1
$\omega 6$	3005	4.3	8.7
$\omega 7$	3035.9	2.8	5.6
$\omega 8$	3075.9	7.1	14.2
$\omega 9$	3166.2	11.9	23.7
$\omega 10$	3384.3	25.3	50.7
n	0.02	0.04	0.08
p	1.3	0.3	0.7
$n1$	0.002	0.001	0.003
R^2	0.96		

Table B.3.24. Fitting parameters for 8 mM C8 Quat with 0% NaCl in H_2O at ppp polarization

Parameters	Estimated Value	Standard Error	Error with 95% Confidence Level
A1	0.3	0.1	0.2
A2	0.3	0.03	0.06
A3	0.6	0.1	0.2
A4	0.9	0.2	0.3
A5	2.7	0.1	0.2
A6	0.02	0.03	0.06
A7	0.1	0.1	0.3
A8	0.01	0.39	0.78
A9	18.2	0.9	1.9
A10	0.3	0.2	0.4
$\Gamma 1$	24.9	8	15.9
$\Gamma 2$	5	0.8	1.6
$\Gamma 3$	7.9	0.9	1.7
$\Gamma 4$	14.9	2.0	4.1
$\Gamma 5$	12.9	0.3	0.6
$\Gamma 6$	0.9	1.5	3.0
$\Gamma 7$	8.1	10.2	20.3
$\Gamma 8$	28.9	2703.1	5406.2
$\Gamma 9$	117.9	4.6	9.2

$\Gamma 10$	20.6	12.7	25.5
$\omega 1$	2739.9	4	8
$\omega 2$	2858	0.5	0.9
$\omega 3$	2889.5	0.5	1.0
$\omega 4$	2910.1	1.2	2.5
$\omega 5$	2959.4	0.3	0.6
$\omega 6$	3015.3	1.2	2.3
$\omega 7$	3074.3	5.4	10.7
$\omega 8$	3117.1	1350.3	2700.6
$\omega 9$	3275.2	3.2	6.4
$\omega 10$	3431.2	7.4	14.7
n	0.003	0.001	0.002
p	0.01	1.7	3.3
n1	0	0	0
R^2	0.98		

Table B.3.25. Fitting parameters for 8 mM C8 Quat with 0.5% NaCl in H_2O at ppp

polarization

Parameters	Estimated Value	Standard Error	Error with 95% Confidence Level
A1	0.2	0.1	0.2
A2	4.1×10^{-07}	0.1	0.2
A3	0.6	0.1	0.2
A4	0.4	0.1	0.2
A5	2.1	0.1	0.2
A6	0.05	0.1	0.1
A7	0.03	0.03	0.07
A8	0.1	0.1	0.2
A9	1.9	0.6	1.3
A10	8.8	0.7	1.4
$\Gamma 1$	35	18.6	37.3
$\Gamma 2$	15.5	3.0×10^{06}	6076256.2
$\Gamma 3$	10.0	1.2	2.5
$\Gamma 4$	12.4	2.3	4.5
$\Gamma 5$	13.3	0.3	0.7

$\Gamma 6$	8.0	11.5	23.1
$\Gamma 7$	3.4	5.3	10.7
$\Gamma 8$	10.0	14.2	28.4
$\Gamma 9$	73.7	16.6	33.2
$\Gamma 10$	97.5	4.9	9.7
$\omega 1$	2710	9.5	19.0
$\omega 2$	2861.1	1.7×10^{06}	3361894.4
$\omega 3$	2890	0.6	1.2
$\omega 4$	2915	1.2	2.4
$\omega 5$	2964.7	0.4	0.8
$\omega 6$	3025	6.5	12.9
$\omega 7$	3078.7	3.4	6.8
$\omega 8$	3104.4	7.6	15.1
$\omega 9$	3327.7	6.3	12.5
$\omega 10$	3500	2.6	5.1
n	0.06	0.01	0.01
p	3.8	0.08	0.16
$n1$	0	0	0
R^2	0.98		

Table B.3.26. Fitting parameters for 8 mM C8 Quat with 1% NaCl in H_2O at ppp polarization

Parameters	Estimated Value	Standard Error	Error with 95% Confidence Level
A1	0.2	0.1	0.1
A2	0	0	0
A3	0.5	0	0.1
A4	1.2	0.1	0.2
A5	3.3	0.1	0.1
A6	0.2	0.1	0.2
A7	0.4	0.1	0.2
A8	0.2	0.1	0.2
A9	3.3	1.3	2.6
A10	11.4	1.3	2.6
$\Gamma 1$	25	6.6	13.2

$\Gamma2$	3.5	0.7	1.4
$\Gamma3$	7.5	0.5	0.9
$\Gamma4$	16.8	1.0	2.0
$\Gamma5$	13.2	0.1	0.3
$\Gamma6$	8.9	3.4	6.8
$\Gamma7$	10.9	3.7	7.4
$\Gamma8$	10.7	8.0	16.0
$\Gamma9$	78.5	13.7	27.3
$\Gamma10$	100	6.8	13.7
$\omega1$	2740	3.8	7.7
$\omega2$	2859	0.4	0.7
$\omega3$	2892	0.2	0.5
$\omega4$	2915	0.5	1.0
$\omega5$	2961.2	0.1	0.3
$\omega6$	3025	2.1	4.2
$\omega7$	3069.2	2.1	4.2
$\omega8$	3100	4.3	8.7
$\omega9$	3350	5.8	11.6
$\omega10$	3468.2	2.9	5.7
n	0.02	0	0
p	3	0.2	0.3
n1	0	0	0
R^2	0.98		

Table B.3.27. Fitting parameters for 8 mM C8 Quat with 5% NaCl in H_2O at ppp

polarization

Parameters	Estimated Value	Standard Error	Error with 95% Confidence Level
A1	0.9	0.2	0.3
A2	0.1	0.1	0.1
A3	1.2	0.1	0.2
A4	0.6	0.2	0.4
A5	3.5	0.2	0.4
A6	1	0.3	0.6
A7	0.3	0.2	0.4

A8	0.3	0.2	0.4
A9	1.7	0.6	1.2
A10	2.9	1.3	2.6
$\Gamma 1$	35	5.8	11.5
$\Gamma 2$	8	4.9	9.9
$\Gamma 3$	12.1	1	2
$\Gamma 4$	12.4	2.8	5.7
$\Gamma 5$	13.8	0.3	0.6
$\Gamma 6$	22.1	5.2	10.3
$\Gamma 7$	10	5.6	11.1
$\Gamma 8$	9.0	6.8	13.7
$\Gamma 9$	47.3	12.9	25.7
$\Gamma 10$	87.3	19.7	39.4
$\omega 1$	2740	4.2	8.3
$\omega 2$	2852	2.5	4.9
$\omega 3$	2890	0.6	1.2
$\omega 4$	2915	1.3	2.6
$\omega 5$	2960.1	0.4	0.7
$\omega 6$	3033.4	2.6	5.3
$\omega 7$	3085.9	3.2	6.4
$\omega 8$	3109.5	3.7	7.4
$\omega 9$	3177.7	6.9	13.9
$\omega 10$	3493.8	9.9	20
n	0.06	0.01	0.03
p	3.7	0.1	0.3
n1	0	0	0
R^2	0.98	0	

Table B.3.28. Fitting parameters for 8 mM C8 Quat with 10% NaCl in H_2O at ppp

polarization

Parameters	Estimated Value	Standard Error	Error with 95% Confidence Level
A1	4.6×10^{-05}	0.1	0.2
A2	0.1	0.1	0.2
A3	0.7	0.1	0.3

A4	1.0	0.3	0.5
A5	2.3	0.1	0.2
A6	0.2	0.4	0.9
A7	1.1	1.0	2.2
A8	0.7	0.6	1.1
A9	0.5	1.5	3.1
A10	7.1	1.1	2.2
$\Gamma 1$	15.6	28824.3	57648.6
$\Gamma 2$	8	4.0	8.0
$\Gamma 3$	10.5	1.2	2.4
$\Gamma 4$	19.5	3.7	7.4
$\Gamma 5$	11.8	0.4	0.7
$\Gamma 6$	11.3	15.9	31.8
$\Gamma 7$	25.6	15.5	31.1
$\Gamma 8$	28.7	17.0	34.1
$\Gamma 9$	70.6	111.1	222.2
$\Gamma 10$	87.6	9.6	19.1
$\omega 1$	2725.4	15526.5	31052.9
$\omega 2$	2857.9	2.1	4.2
$\omega 3$	2892	0.7	1.4
$\omega 4$	2915	2.0	4.1
$\omega 5$	2959	0.3	0.6
$\omega 6$	3024.9	5.3	10.6
$\omega 7$	3050.4	7.0	14.1
$\omega 8$	3100	10.4	20.9
$\omega 9$	3349.6	47.8	95.7
$\omega 10$	3466	4.8	9.7
n	0.02	0.01	0.02
p	4.1	0.1	0.2
n1	0.0	0.0	0.0
R^2	0.98		

Table B.3.29. Fitting parameters for 8 mM C10 Quat with 0% NaCl in H_2O at ppp polarization

Parameters	Estimated Value	Standard Error	Error with 95% Confidence Level
A1	0.7	0.1	0.2
A2	0.01	0.08	0.15
A3	1.0	0.1	0.2
A4	0.3	0.1	0.2
A5	3.6	0.1	0.1
A6	0.01	0.19	0.38
A7	0.0	0.2	0.4
A8	0.0	0.1	0.1
A9	10.0	0.8	1.7
A10	8.1	0.5	1.0
$\Gamma1$	24.9	3.2	6.3
$\Gamma2$	13.5	86.3	172.7
$\Gamma3$	12.9	1.3	2.6
$\Gamma4$	9.6	1.8	3.6
$\Gamma5$	10.8	0.2	0.4
$\Gamma6$	12.2	334.1	668.2
$\Gamma7$	14.4	1249.0	2498.0
$\Gamma8$	5.8	265.3	530.7
$\Gamma9$	75.2	3.7	7.4
$\Gamma10$	69.9	3.5	7.1
$\omega1$	2738.1	2.6	5.2
$\omega2$	2855.4	42.5	85.0
$\omega3$	2890	0.6	1.3
$\omega4$	2915	1.0	2.0
$\omega5$	2961.7	0.2	0.5
$\omega6$	3013.5	217.0	434.1
$\omega7$	3053.8	741.4	1482.7
$\omega8$	3117.3	179.6	359.3
$\omega9$	3289.8	1.6	3.1
$\omega10$	3447.2	1.9	3.9
n	0.02	0.01	0.02
p	3.4	0.2	0.4
n1	0.0	0.0	0.0
R^2	0.99		

Table B.3.30. Fitting parameters for 8 mM C10 Quat with 1% NaCl in H_2O at ppp polarization

Parameters	Estimated Value	Standard Error	Error with 95% Confidence Level
A1	0.3	0.04	0.09
A2	0.03	0.01	0.02
A3	0.4	0.02	0.04
A4	0.3	0.05	0.1
A5	2.4	0.1	0.2
A6	0.1	0.06	0.1
A7	0.3	0.1	0.2
A8	0.1	0.1	0.2
A9	4.0	0.9	1.7
A10	6.2	0.4	0.9
$\Gamma1$	25.0	2.7	5.4
$\Gamma2$	4.0	2.0	4.0
$\Gamma3$	6.9	0.4	0.8
$\Gamma4$	13.3	1.6	3.1
$\Gamma5$	11.9	0.1	0.3
$\Gamma6$	11.7	5.3	10.7
$\Gamma7$	19.6	5.9	11.8
$\Gamma8$	20.0	17.2	34.3
$\Gamma9$	95.7	10.0	20.1
$\Gamma10$	78.5	3.7	7.5
$\omega1$	2710	1.8	3.7
$\omega2$	2854.5	1.4	2.9
$\omega3$	2888	0.3	0.6
$\omega4$	2915	0.7	1.5
$\omega5$	2962.1	0.1	0.2
$\omega6$	3029.9	2.8	5.6
$\omega7$	3076.6	2.9	5.8
$\omega8$	3130	9.0	18.1
$\omega9$	3300	4.6	9.3
$\omega10$	3451	2.0	3.9
n	0.08	0.01	0.02

p	4.1	0.04	0.09
n1	0	0	0
R^2	0.99		

Table B.3.31. Fitting parameters for 8 mM C10 Quat with 10% NaCl in H_2O at ppp

polarization

Parameters	Estimated Value	Standard Error	Error with 95% Confidence Level
A1	0.0	0.06	0.10
A2	0.01	0.01	0.01
A3	1.5	0.2	0.4
A4	1.2	0.3	0.6
A5	1.9	0.1	0.2
A6	1.3	0.4	0.8
A7	0.9	0.8	1.6
A8	0.3	0.7	1.4
A9	0.5	0.2	0.4
A10	1.3	0.3	0.7
$\Gamma1$	15.3	5025.7	10051.4
$\Gamma2$	1.0	0.8	1.5
$\Gamma3$	17.0	1.3	2.6
$\Gamma4$	20.0	3.3	6.5
$\Gamma5$	12.7	0.5	0.9
$\Gamma6$	20.0	3.6	7.3
$\Gamma7$	20.0	11.8	23.6
$\Gamma8$	20.0	25.7	51.4
$\Gamma9$	15.6	7.4	14.8
$\Gamma10$	40.0	11.5	23.0
$\omega1$	2732.6	2739.9	5479.8
$\omega2$	2851.2	0.5	0.9
$\omega3$	2885.0	0.8	1.6
$\omega4$	2915.3	1.4	2.8
$\omega5$	2955.0	0.3	0.6
$\omega6$	3028.2	2.0	3.9
$\omega7$	3071.2	5.7	11.4

$\omega 8$	3100.4	17.3	34.7
$\omega 9$	3250.0	5.1	10.2
$\omega 10$	3451.2	9.7	19.3
n	0.01	0.0	0.0
p	4.04	0.3	0.6
n1	0.0	0.0	0.0
R^2	0.98	0	

Table B.3.32. Fitting parameters for 8 mM C12 Quat with 0% NaCl in H_2O at ppp polarization

Parameters	Estimated Value	Standard Error	Error with 95% Confidence Level
A1	0.4	0.1	0.2
A2	0.1	0.04	0.08
A3	0.6	0.1	0.2
A4	1.0	0.1	0.3
A5	2.7	0.1	0.2
A6	0.1	0.1	0.2
A7	0.8	0.2	0.3
A8	1.3	1.2	2.4
A9	15.0	2.3	4.5
A10	8.6	1.1	2.1
$\Gamma 1$	25.0	6.1	12.2
$\Gamma 2$	6.0	2.6	5.2
$\Gamma 3$	8.8	1.0	2.0
$\Gamma 4$	15.1	1.8	3.7
$\Gamma 5$	11.4	0.3	0.5
$\Gamma 6$	10.3	10.1	20.3
$\Gamma 7$	14.7	3.0	6.0
$\Gamma 8$	60.0	37.1	74.2
$\Gamma 9$	100.4	9.3	18.5
$\Gamma 10$	79.4	5.8	11.6
$\omega 1$	2710.0	3.8	7.5
$\omega 2$	2858.5	1.5	3.0

$\omega 3$	2886	0.5	1.0
$\omega 4$	2910	1.0	1.9
$\omega 5$	2960.4	0.4	0.7
$\omega 6$	3010	5.2	10.4
$\omega 7$	3064.6	1.4	2.9
$\omega 8$	3150.0	12.8	25.6
$\omega 9$	3277.6	3.1	6.2
$\omega 10$	3437.5	2.7	5.5
n	0.01	0.001	0.002
p	1.9×10^{-07}	0.5	1.0
n1	0	0	0
R^2	0.98		

Table B.3.33. Fitting parameters for 8 mM C12 Quat with 1% NaCl in H_2O at ppp

polarization

Parameters	Estimated Value	Standard Error	Error with 95% Confidence Level
A1	0.4	0.08	0.1
A2	0.06	0.01	0.02
A3	0.9	0.1	0.2
A4	0.7	0.2	0.3
A5	1.7	0.07	0.1
A6	0.0	0.1	0.3
A7	0.5	0.2	0.3
A8	0.1	0.3	0.7
A9	9.9	6.3	12.8
A10	10.0	4.9	9.8
$\Gamma 1$	25.0	5.9	11.7
$\Gamma 2$	2.7	0.9	1.9
$\Gamma 3$	11.4	1.0	2.0
$\Gamma 4$	16.1	3.2	6.3
$\Gamma 5$	11.0	0.3	0.6
$\Gamma 6$	9.4	3735.7	7471.3
$\Gamma 7$	11.8	4.1	8.2

$\Gamma 8$	24.7	49.6	99.2
$\Gamma 9$	129.9	38.3	76.6
$\Gamma 10$	118.2	22.4	44.8
$\omega 1$	2740.0	3.8	7.6
$\omega 2$	2862.7	0.7	1.3
$\omega 3$	2894.1	0.6	1.2
$\omega 4$	2917.9	1.6	3.2
$\omega 5$	2964.6	0.3	0.6
$\omega 6$	3038.7	1823.3	3646.6
$\omega 7$	3065.4	1.9	3.8
$\omega 8$	3129.9	21.1	42.1
$\omega 9$	3261.5	14.0	28.1
$\omega 10$	3404.0	9.2	18.5
n	0.01	0.005	0.01
p	1.7	0.2	0.4
$n1$	0.0	0.0	0.0
R^2	0.98		

Table B.3.34. Fitting parameters for 8 mM C12 Quat with 10% NaCl in H_2O at ppp polarization

Parameters	Estimated Value	Standard Error	Error with 95% Confidence Level
A1	7.1×10^{-08}	0.08	0.2
A2	7.2×10^{-08}	0.02	0.04
A3	0.7	0.2	0.4
A4	1.2	0.6	1.1
A5	2.8	0.2	0.4
A6	1.0	0.7	1.3
A7	1.3	3.9	7.8
A8	3.7	5.7	11.4
A9	4.9	2.8	5.6
A10	10.0	2.2	4.5
$\Gamma 1$	13.4	1.7×10^{-08}	35226001.3
$\Gamma 2$	1.8	686708.3	1373416.6
$\Gamma 3$	8.8	1.3	2.5

$\Gamma 4$	25.0	8.7	17.4
$\Gamma 5$	14.4	0.7	1.5
$\Gamma 6$	20.4	8.2	16.4
$\Gamma 7$	36.1	49.5	99.1
$\Gamma 8$	64.5	43.2	86.3
$\Gamma 9$	80.0	26.5	52.9
$\Gamma 10$	100.2	12.8	25.6
$\omega 1$	2725.3	1.2×10^{07}	24354788.6
$\omega 2$	2861.3	450966.4	901932.8
$\omega 3$	2894.3	0.7	1.5
$\omega 4$	2920	3.5	7.1
$\omega 5$	2965.8	0.6	1.2
$\omega 6$	3028.8	3.5	7.1
$\omega 7$	3079.3	14.8	29.5
$\omega 8$	3130	38.0	76.1
$\omega 9$	3280	8.3	16.5
$\omega 10$	3430	7.2	14.3
n	0.03	0.008	0.02
p	1.1	0.09	0.2
n1	0.0	0.0	0.0
R^2	0.97	0	

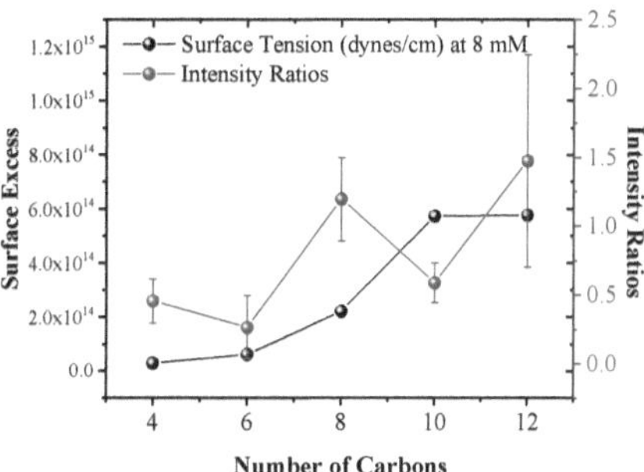

Figure B.4. The plot of surface excess and intensity ratio of CH_2 SS and CH_3 SS vs number of carbons in the alkyl chain.

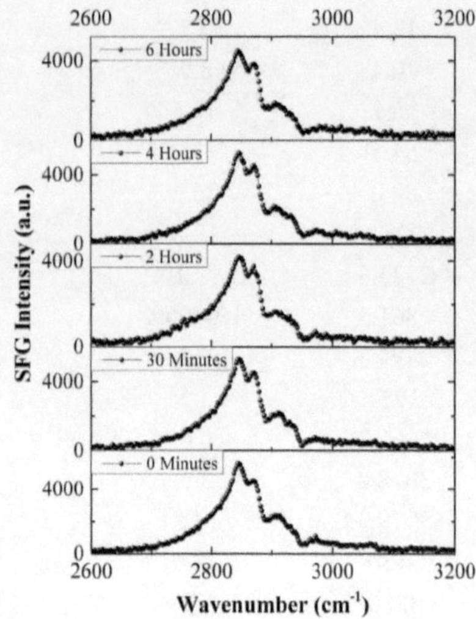

Figure B.5. SFG spectra of 8 mM C8 in water at 0, 0.5, 2, 4, and 6 hours at ssp

polarization combinations.

Table B.4. The pH values of 8mM Quat solution in water.

Quats	pH
C4	5.28±0.01
C6	5.51±0.04
C8	5.36±0.07
C10	5.35±0.07
C12	5.21±0.02

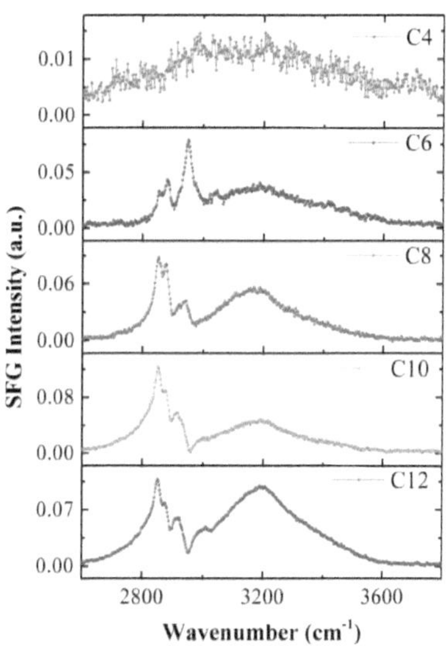

Figure B.6. SFG spectra of 1mM C4, C6, C8, C10, and C12 alkyltetramethylammonium

bromide (ATAB) compounds in water (without NaCl) at ssp polarization combinations.

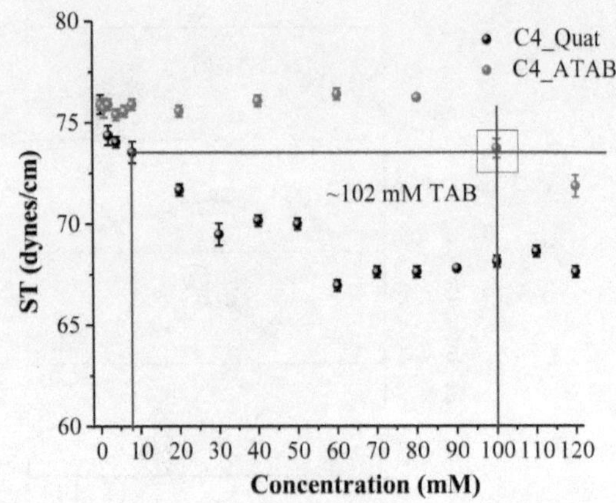

Figure B.7. Surface tension of C4 Quat and C4 TAB (Without salt).

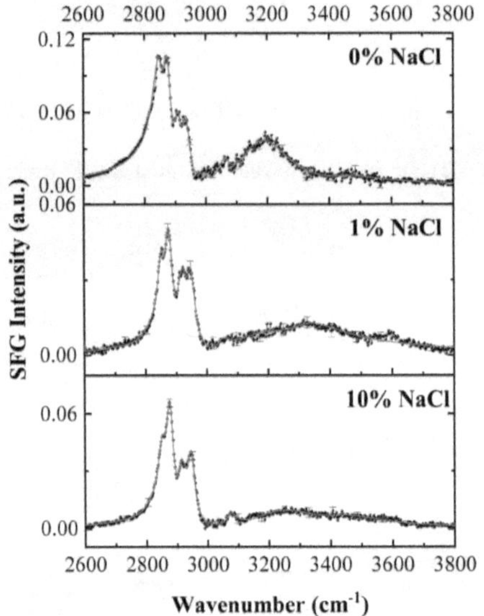

Figure B.8. Fitted SFG spectra of 8 mM C4 in water (0, 1, and 10 % NaCl) at ssp polarization combinations.

Table B.5. Ionic strength of 8mM Quats in water with 0%, 0.5%, 1%, 5%, 10% NaCl.

Quats Concentration (M)	NaCl Concentration (M)	Ionic strength (M)
0.008	0	0.008
0.008	0.085	0.093
0.008	0.17	0.178
0.008	0.850	0.858
0.008	1.7	1.708

Table B.6. The parameters for generating the simulated SFG curves for the orientational analysis of Quats.

Parameters	
Refractive indices: n_1(air), n_2(liquid) and n_i(interface)	$n_{1,SFG} = n_{1,vis} = n_{1,IR} = 1.0$
	$n_{2,SFG} = 1.3312$
	$n_{2,vis} = 1.3292$
	$n_{2,IR} = 1.4260$
	$n_{i,SFG} = n_{i,vis} = n_{i,IR} = 1.1656$
R-value	$3.4(-CH_3)$
$\beta_{a,c,a}$	3.4
$\beta_{a,a,c} = \beta_{b,b,c} = \beta_{c,c,c}$	1.0
N_s(number density)	1.0
$\omega_{vis}(cm^{-1})$	12578.6
$\omega_{IR}(cm^{-1})$	2900
$\omega_{SFG}(cm^{-1})$	15478.6
$\theta_{vis}(°)$	50
$\theta_{IR}(°)$	60
$\theta_{SFG}(°)$	51.7
$L_{xx,SFG}$	0.6132
$L_{xx,vis}$	0.6284
$L_{xx,IR}$	0.5207
$L_{yy,SFG}$	0.7313
$L_{yy,vis}$	0.7435

$L_{yy,IR}$	0.6124
$L_{zz,SFG}$	0.5837
$L_{zz,vis}$	0.5764
$L_{zz,IR}$	0.6029
Example: $\chi^{(2)}_{eff,ssp,ss}$	0.163926 (4.4 Cos[θ]+2.4 Cos[θ]^3)

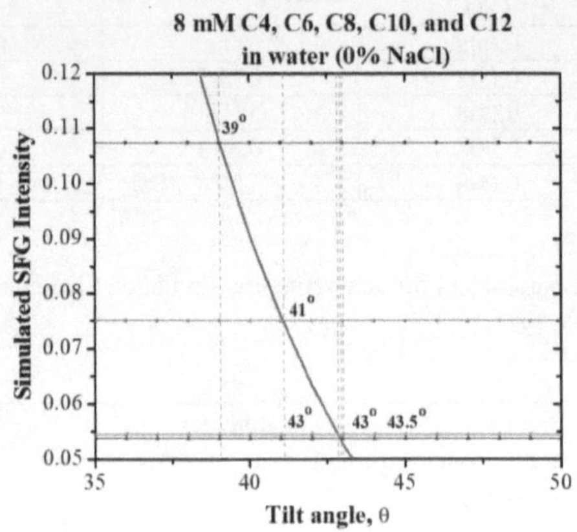

Figure B.9. The simulated curves of the SFG intensity ratios (CH₃ SS of PPP/CH₃ AS PPP) as a function of tilt angle of C4 = 39°, C6 = 43.5°, C8 = 43°, C10 = 41°, and C12 = 43°.

Table B.7. Tilt angle values for Quat compounds and are reported with 95% confidence level.

Quats	0% NaCl	1% NaCl	10% NaCl
	Tilt angle, θ	Tilt angle, θ	Tilt angle, θ
C4	39° ± 27°	45° ± 84°	52.5° ± 58°
C6	43.5° ± 58°	62° ± 2151	49° ± 219°

C8	43° ± 27°	47° ± 15°	40° ± 34°
C10	41° ± 18°	47° ± 13°	28° ± 15°
C12	43° ± 25°	34° ±19°	41.5° ± 41°

Table B.8. The fitted R^2 values of 8mM Quats in water with 0%, 1%, and 10% NaCl at ssp and ppp polarization combinations.

Quats	ssp			ppp		
	R^2			R^2		
	0% NaCl	1% NaCl	10% NaCl	0% NaCl	1% NaCl	10% NaCl
C4	0.99	0.99	0.99	0.99	0.99	0.98
C6	0.97	0.99	0.98	0.97	0.98	0.96
C8	0.98	0.99	0.99	0.98	0.98	0.98
C10	0.97	0.97	0.98	0.99	0.99	0.98
C12	0.97	0.99	0.99	0.98	0.98	0.97

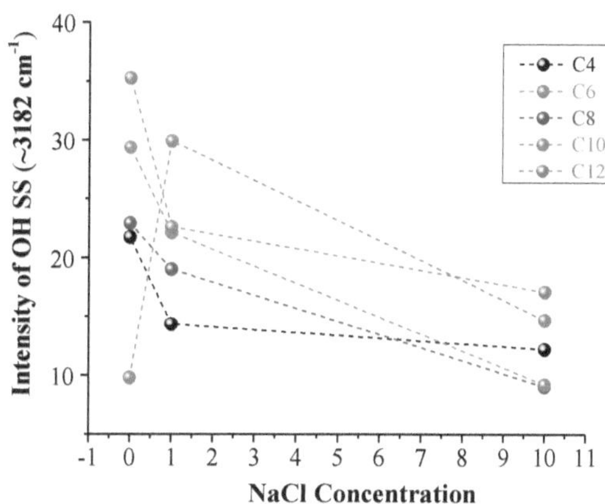

Figure B.10. The intensity of OH SS at ~3182 cm^{-1} as a function of NaCl concentration

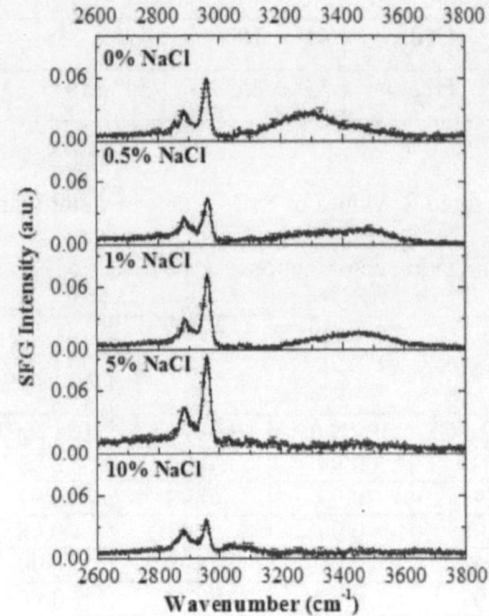

Figure B.11. Fitted SFG spectra of 8mM C8 in water with 0%, 0.5%, 1%, 5%, 10% NaCl at ppp polarization combinations.

APPENDIX C: SUPPORTING DETAILS FOR CHAPTER 5

Table C1: Fitting parameters for C4 Quat in D_2O at liquid-solid interface.[24, 49, 50]

Parametes	SSP		Parameters	PPP	
	Estimated Value	Standard Error		Estimated Value	Standard Error
A1	50	36.2	**A1**	78.8	6.7
A2	50	32.5	**A2**	195	9.7
A3	174.5	9.9	**A3**	5.2×10^{-8}	44.87
A4	20	5.3	**A4**	60	5.7
F1	13.3	9.5	**F1**	7.6	0.8
F2	10	3.8	**F2**	8	0.3
F3	7.5	0.2	**F3**	59.4	1.1×10^9
F4	7.9	1.7	**F4**	5	0.6
$\omega 1$	2865	6.2	$\omega 1$	2873.1	0.5
$\omega 2$	2885	2.1	$\omega 2$	2916.2	0.2
$\omega 3$	2907.4	0.1	$\omega 3$	2950.5	0.006
$\omega 4$	2951	0.9	$\omega 4$	2972	3.8×10^9
n	6.2	0.4	**n**	8.2	0.8
P	-0.05	0.1	**P**	4.4	0.04
n1	176.1`	3.2	**n1**	272.1	9.0
$\mathbf{R^2}$	0.96		$\mathbf{R^2}$	0.96	

Table C2: Fitting parameters for C12 Quat in D_2O at liquid-solid interface.[24, 49, 50]

Parameters	SSP		Parameters	PPP	
	Estimated Value	Standard Error		Estimated Value	Standard Error
A1	1.5×10^{-6}	1.6	**A1**	229.2	0.1
A2	210.9	0.2	**A2**	0	0.2
A3	1.4×10^{-6}	1.9	**A3**	19.5	0.07
A4	131.4	0.1	**A4**	144.3	0.1
F1	61.9	3.6×10^6	**F1**	6.8	0.002
F2	8.9	0.007	**F2**	9.9	1.6×10^{-16}
F3	74.1	2.5×10^6	**F3**	1.9	0.009
F4	7.7	0.007	**F4**	5.5	0.003
$\omega 1$	2859.9	2.9×10^7	$\omega 1$	2882	0.002
$\omega 2$	2882.4	0.005	$\omega 2$	2920	2.7×10^{-17}
$\omega 3$	2909.9	3.1×10^7	$\omega 3$	2943.9	0.006
$\omega 4$	2945.5	0.006	$\omega 4$	2968.5	0.003

n	10.20	0.01	**n**	4.6	0.01
P	-0.9	0.001	**P**	-1.1	0.001
n1	188.1`	0.15	**n1**	282.6	0.0997
R^2	0.96		**R^2**	0.94	

Simulation Details

The molecular dynamics (MD) simulation system comprised of 42 surfactant molecules, their counter-ions, and 1502 water molecules near a metal slab. The simulation box was 34.62 Å x 34.98 Å x 80 Å in dimensions and was periodic in the X and Y directions. The Z-direction was non-periodic because of the presence of a metal slab at one side (spanning from Z = 0 Å to Z = 11.78 Å) of the simulation box. The metal slab comprised of six layers of gold atoms arranged in (111) planes of face-centered cubic (fcc) crystal structure, with a lattice constant of 4.08 Å. The aqueous medium extended up to ~60 Å in the Z direction. Simulations were performed at a temperature, T = 300 K in the isothermal-isochoric (NVT) ensemble. A vacuum space of ~20 Å was kept beyond the aqueous column so that the system pressure is maintained at the equilibrium vapor pressure at T = 300 K.[151] The face of the simulation box opposite of the metal slab had an athermal surface to prevent molecules from leaving the box.

Partial charges on the surfactant molecules were determined by performing density functional theory (DFT) calculations on the Gaussian 16 program.[152] DFT calculations were performed by employing B3LYP hybrid functional with 6-31G(d,p) basis set and water as the implicit solvent. Interactions of surfactant molecules were modeled using the general amber force field (GAFF), which is a widely used force field for organic molecules.[153] The quat molecules are protonated and carry a charge of +1. Bromide ions were introduced as the counter-ions to keep the entire system charge neutral. Interaction parameters of bromide ions were obtained from the Joung-Cheatham's model.[154] Water was modeled via the extended simple point charge (SPC/E) model.[155] The parameters for gold atoms were obtained from the interface force field developed by Heinz and co-workers, a force field that is compatible with GAFF.[156] The Lennard-Jones and short-range Coulombic interactions were cut-off at a distance of 10 Å. Long-range Coulombic interactions were computed using the particle-particle particle-mesh (PPPM) method. All MD simulations were performed using the Large-scale Atomic/Molecular Massively Parallel Simulator (LAMMPS).[157]

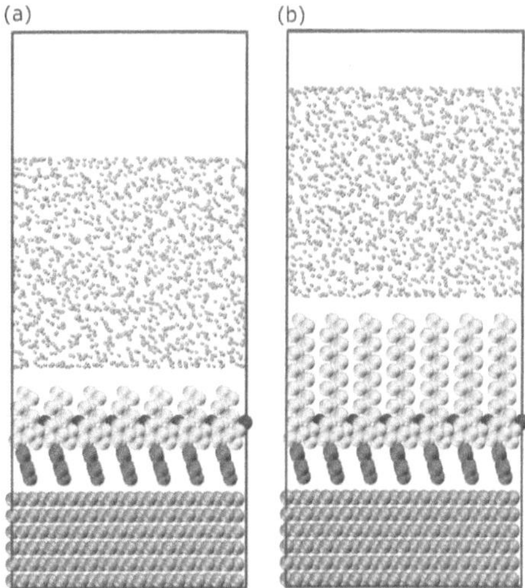

Figure C3. Snapshots of configuration of (a) C4 molecules, and (b) C12 molecules at t = 0 ns.

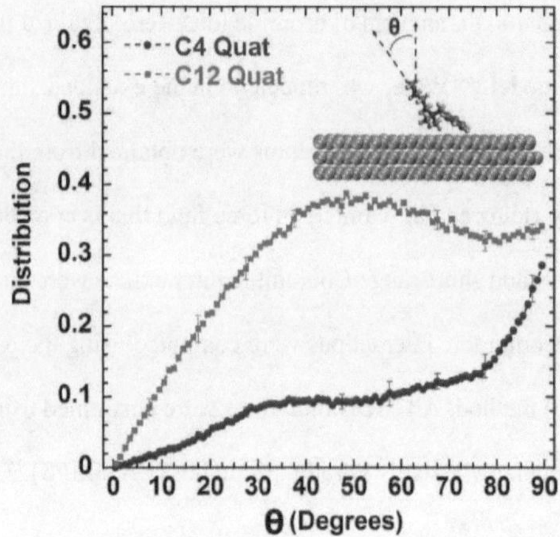

Figure C4. Tilt angle distribution of terminal methyl groups with respect to the surface normal. The angle is measured between the terminal CH_2-CH_3 vector and the surface normal. The distribution is plotted for the adsorbed molecules whose tilt angle lies within 90° with the surface normal. The average value of tilt angles was found to be 59.3° ± 1.4° for C4 Quat molecules and 52.1° ± 1.0° for C12 Quat molecules.

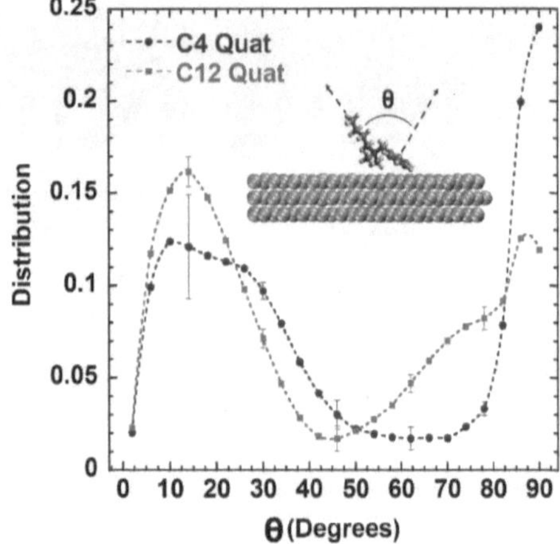

Figure C5. Distribution for intra-molecular orientation of alkyl tails with respect to the normal vector of the aromatic rings of adsorbed C4 and C12 Quat molecules. A value close to 90° implies that the molecule is not bent with respect to its bulk phase conformation (**Figure C6**). A broad distribution from 0° to 45° in this **Figure** shows that a good fraction of the adsorbed molecules are bent such that their alkyl tails are not aligned parallel to the plane of their aromatic rings.

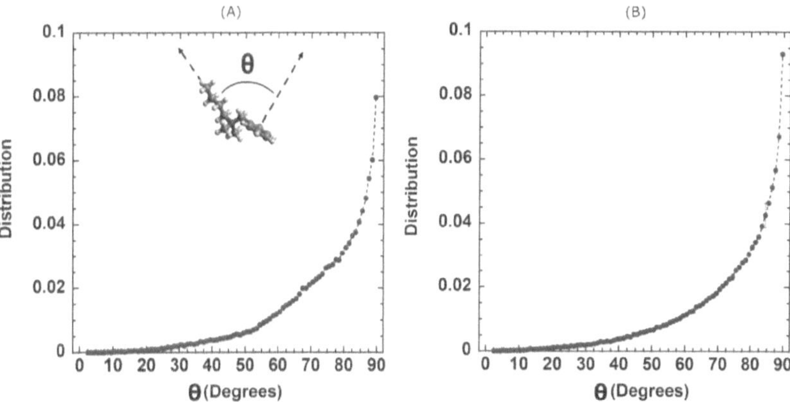

Figure C6. Distribution for intra-molecular orientation of alkyl tails with respect to the normal vector to the aromatic rings of (A) C4 molecules, and (B) C12 molecules in the bulk aqueous phase. A peak close to 90° shows that the molecules in the bulk aqueous phase have their alkyl tails aligned parallel to the plane of their aromatic rings. Some error bars are smaller than the size of the markers.

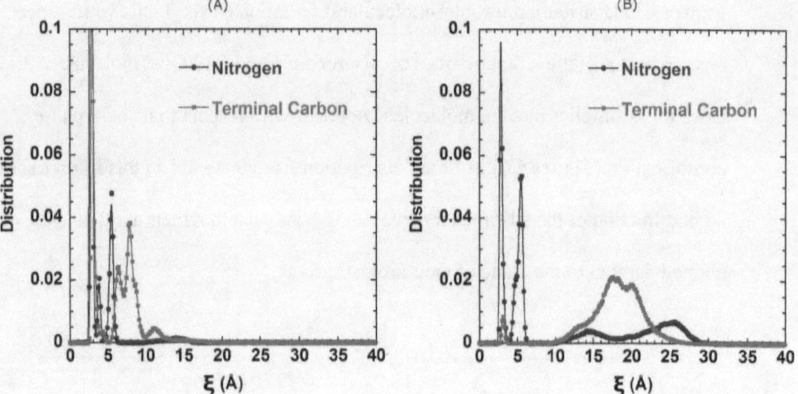

Figure C7. Distribution of nitrogens and terminal carbons (carbon atom of the terminal CH$_3$ groups) of adsorbed (A) C4 molecules, and (B) C12 molecules as a function of distance from the metal surface, ξ. The location of the peaks shows that unlike C4 molecules, C12 molecules mostly stand-up on the surface.

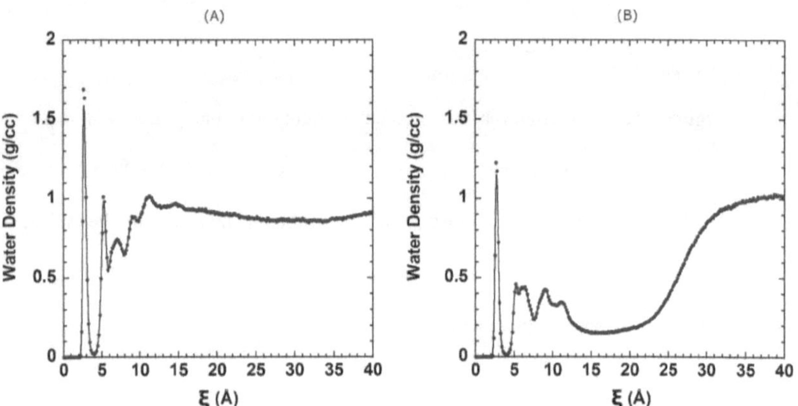

Figure C8. Density profile of water as a function of distance from the gold surface, ξ in presence of adsorbed (A) C4 molecules, and (B) C12 molecules.

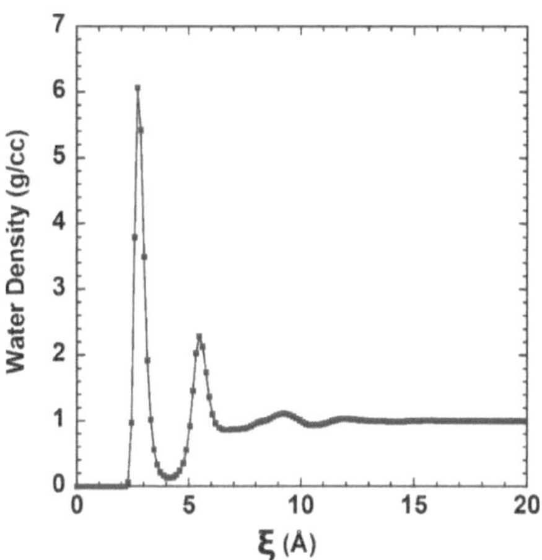

Figure C9. Density profile of water as a function of distance from the gold surface, ξ in absence of surfactant molecules in the system.

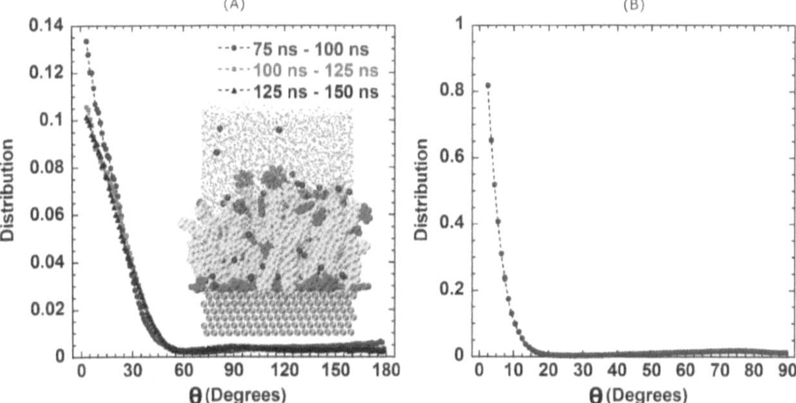

Figure C10. Distribution of orientation of (A) alkyl tails, and (B) normal vector of the aromatic rings with respect to the surface normal of adsorbed C12 molecules, in the simulation box of size 49.05 Å × 49.97 Å × 80 Å consisting of 81 C12 molecules.

APPENDIX D: COPYRIGHT PERMISSION DOCUMENTS

Role of the cationic headgroup to conformational changes undergone by shorter alkyl chain surfactant and water molecules at the air-liquid interface

Author: Md. Rubel Khan,Uvinduni I. Premadasa,Katherine Leslee A. Cimatu

Publication: Journal of Colloid and Interface Science

Publisher: Elsevier

Date: 15 May 2020

Journal Author Rights

Please note that, as the author of this Elsevier article, you retain the right to include it in a thesis or dissertation, provided it is not published commercially. Permission is not required, but please ensure that you reference the journal as the original source. For more information on this and on your other retained rights, please visit: https://www.elsevier.com/about/our-business/policies/copyright#Author-rights

BACK   CLOSE WINDOW

Home Help Email Support Sign in Create Account

Direct Observation of Adsorption Morphologies of Cationic Surfactants at the Gold Metal–Liquid Interface

Author: Md. Rubel Khan, Himanshu Singh, Sumit Sharma, et al

Publication: Journal of Physical Chemistry Letters

Publisher: American Chemical Society

Date: Nov 1, 2020

Copyright © 2020, American Chemical Society

PERMISSION/LICENSE IS GRANTED FOR YOUR ORDER AT NO CHARGE

This type of permission/license, instead of the standard Terms & Conditions, is sent to you because no fee is being charged for your order. Please note the following:

- Permission is granted for your request in both print and electronic formats, and translations.
- If figures and/or tables were requested, they may be adapted or used in part.
- Please print this page for your records and send a copy of it to your publisher/graduate school.
- Appropriate credit for the requested material should be given as follows: "Reprinted (adapted) with permission from {COMPLETE REFERENCE CITATION}. Copyright {YEAR} American Chemical Society." Insert appropriate information in place of the capitalized words.
- One-time permission is granted only for the use specified in your request. No additional uses are granted (such as derivative works or other editions). For any other uses, please submit a new request.

BACK   CLOSE WINDOW